T0220217

Advanced Object-Oriented Programming in R

Statistical Programming for Data Science, Analysis and Finance

Thomas Mailund

Apress®

Advanced Object-Oriented Programming in R: Statistical Programming for Data Science, Analysis and Finance

Thomas Mailund
Aarhus N, Denmark

ISBN-13 (pbk): 978-1-4842-2918-7   ISBN-13 (electronic): 978-1-4842-2919-4
DOI 10.1007/978-1-4842-2919-4

Library of Congress Control Number: 2017945396

Cover image by Freepik (`www.freepik.com`)

Managing Director: Welmoed Spahr
Editorial Director: Todd Green
Acquisitions Editor: Steve Anglin
Development Editor: Matthew Moodie
Technical Reviewer: Karthik Ramasubramanian
Coordinating Editor: Mark Powers
Copy Editor: Larissa Shmailo

Distributed to the book trade worldwide by Springer Science+Business Media New York, 233 Spring Street, 6th Floor, New York, NY 10013. Phone 1-800-SPRINGER, fax (201) 348-4505, e-mail `orders-ny@springer-sbm.com`, or visit `www.springeronline.com`. Apress Media, LLC is a California LLC and the sole member (owner) is Springer Science + Business Media Finance Inc (SSBM Finance Inc). SSBM Finance Inc is a **Delaware** corporation.

For information on translations, please e-mail `rights@apress.com`, or visit `http://www.apress.com/rights-permissions`.

Apress titles may be purchased in bulk for academic, corporate, or promotional use. eBook versions and licenses are also available for most titles. For more information, reference our Print and eBook Bulk Sales web page at `http://www.apress.com/bulk-sales`.

Any source code or other supplementary material referenced by the author in this book is available to readers on GitHub via the book's product page, located at `www.apress.com/9781484229187`. For more detailed information, please visit `http://www.apress.com/source-code`.

Printed on acid-free paper

Contents at a Glance

About the Author ix

About the Technical Reviewer xi

Introduction xiii

■Chapter 1: Classes and Generic Functions 1

■Chapter 2: Class Hierarchies 21

■Chapter 3: Implementation Reuse 35

■Chapter 4: Statistical Models 43

■Chapter 5: Operator Overloading 61

■Chapter 6: S4 Classes 73

■Chapter 7: R6 Classes 91

■Chapter 8: Conclusions 107

Index... 109

Contents

About the Author ix

About the Technical Reviewer xi

Introduction xiii

■Chapter 1: Classes and Generic Functions 1

Generic Functions 1

Classes 3

Polymorphism in Action 5

Designing Interfaces 9

The Usefulness of Polymorphism 12

Polymorphism and Algorithmic Programming 13

Sorting Lists 14

General Comments on Flexible Implementations of Algorithms 18

More on UseMethod 19

■Chapter 2: Class Hierarchies 21

Interfaces and Implementations 21

Polymorphism and Interfaces 22

Abstract and Concrete Classes 23

Implementing Abstract and Concrete Classes in R 24

Another Example: Graphical Objects 25

Class Hierarchies As Interfaces with Refinements 30

Chapter 3: Implementation Reuse 35

Method Lookup in Class Hierarchies 36

Getting the Hierarchy Correct in the Constructors 38

NextMethod 39

Chapter 4: Statistical Models 43

Bayesian Linear Regression 43

Model Matrices 47

Constructing Fitted Model Objects 52

Coefficients and Confidence Intervals 53

Predicting Response Variables 54

Chapter 5: Operator Overloading 61

Functions and Operators 62

Defining Single Operators 63

Group Operators 64

Units Example 66

Chapter 6: S4 Classes 73

Defining S4 Classes 73

Generic Functions 75

Slot Prototypes 76

Object Validity 77

Generic Functions and Class Hierarchies 78

Requiring Methods 82

Constructors 83

Dispatching on Type-Signatures 84

Operator Overloading 86

Combining S3 and S4 Classes 88

Chapter 7: R6 Classes **91**

Defining Classes 91

    Object Initialization 94

    Private and Public Attributes 95

    Active Bindings 97

Inheritance 98

References to Objects and Object Sharing 99

Interaction with S3 and Operator Overloading 103

Chapter 8: Conclusions **107**

Index **109**

Chapter 7: The Classes ....

Defining Classes ....

Object Initialization ....

Private and Public Virtual ....

Active Elements ....

Inheritance ....

Reference to Objects and Object Sharing ....

Interaction with $5 and Other for Overloading ....

Another Bank Example ....

About the Author

Thomas Mailund is an associate professor in bioinformatics at Aarhus University, Denmark. He has a background in math and computer science. For the last decade, his main focus has been on genetics and evolutionary studies, particularly comparative genomics, speciation, and gene flow between emerging species. He has published *Beginning Data Science in R*, *Functional Programming in R* and *Metaprogramming in R* with Apress, as well as other books out there.

About the Technical Reviewer

Karthik Ramasubramanian works for one of the largest and fastest-growing technology unicorns in India, Hike Messenger. He brings the best of business analytics and data science experience to his role at Hike Messenger. In his seven years of research and industry experience, he has worked on cross-industry data science problems in retail, e-commerce, and technology, developing and prototyping data-driven solutions. In his previous role at Snapdeal, one of the largest e-commerce retailers in India, he was leading core statistical modeling initiatives for customer growth and pricing analytics. Prior to Snapdeal, he was part of a central database team, managing the data warehouses for global business applications of Reckitt Benckiser (RB). He has vast experience working with scalable machine learning solutions for industry, including sophisticated graph network and self-learning neural networks. He has a Master's in theoretical computer science from PSG College of Technology, Anna University, and is a certified big data professional. He is passionate about teaching and mentoring future data scientists through different online and public forums. He enjoys writing poems in his leisure time and is an avid traveler.

Introduction

Welcome to *Object-oriented Programming in R*. I wrote this book to have teaching material beyond the typical introductory level of most textbooks on R. This book is intended to introduce objects and classes in R and how object-oriented programming is done in R. Object-oriented programming is based on the concept of *objects* and on designing programs in terms of operations that one can do with objects and how objects communicate with other objects.

This is often thought of in terms of objects with states, where operations on objects change the object state. Think of an object such as a bank account. Its state would be the amount on it, and inserting or withdrawing money from it would change its state. Operations we do on objects are often called "methods" in the literature, but in some programming languages the conceptual model is that objects are communicating and sending each other messages, and the operations you do on an object are how it responds to messages it receives.

In R, data is immutable, so you don't write code where you change an object's state. Rather, you work with objects as values, and operations on objects create new objects when you need new "state". Objects and classes in R are more like abstract data structures. You have values and associated operations you can do on these values. Such abstract data structures are implemented differently in different programming languages. Most object-oriented languages implement them using classes and class hierarchies while many functional languages define them using some kind of *type specifications* that define which functions can be applied to objects.

Types determine what you can do with objects. You can, for example, add numbers, and you can concatenate strings, but you can't really add strings or concatenate numbers. In some programming languages, so-called statically typed languages, you associate types with variables, which restrict which objects the variables can refer to and enables some consistency check of code before you run it. In such languages, you can specify new types by defining which operations you can do on them, and you then need to add type specifications to variables referring to them. Other programming languages, called *dynamically typed* languages, do not associate types with variables but let them refer to any kind of objects. R is dynamically typed, so you do not specify abstract data types through a type specification. The operations you can do on objects are simply determined by which functions you can call on the objects. You can still think of these as specifications of abstract data structures; however, they are just implicitly defined.

Abstract data structures can be implemented in different ways, which is what makes them *abstract*, and the way to separate implementation from an interface is through *polymorphic* or *generic* functions, a construction founded on object-oriented programming. Generic functions are implemented through a *class* mechanism, also derived from object-oriented programming. The functions implemented by a class determine the interface of objects in the class, and by constructing hierarchies of classes, you can share the implementation of common functions between classes.

Abstract data structures are often used in algorithmic programming to achieve efficient code, but such programming is frequently not the objective of R programs. There, we are more interested in fitting data to models and such, which frequently does not require algorithmic data structures. Fitted models, however, are also examples of abstract data structures in the sense that I use the term in this book. Models have an abstract interface that allows us to plot fitted models, predict new response variables for new data, and so forth, and we can use the same generic functions for such operations. Different models implement their own versions of these generic functions, so you can write generic code that will work on linear models, decision trees, or neural networks, for example.

Object-oriented programming was not built into the R language initially but was added later, and unfortunately, more than one object-oriented system was added. There are actually three different ways to implement object-oriented constructions in R, each with different pros and cons, and these three systems do not operate well together. I will cover all three in this book (S3, S4, and R6) but put most emphasis on the S3 system which is the basis of the so-called "tidy verse", the packages such as `tidyr`, `dplyr`, `ggplot2`, etc., which form the basis of most data analysis pipelines these days.

When developing your own software, I will strongly recommend that you stick to one object-oriented system instead of mixing them, but which one you choose is a matter of taste and which other packages your code is intended to work with.

Most books I have read on object-oriented programming, and the classes I have taken on object-oriented programming, have centered on object-oriented modeling and software design. There, the focus is on how object-orientation can be used to structure how you think about your software and how the software can reflect physical or conceptual aspects of the world that you try to model in your software. If, for instance, you implement software for dealing with accounting, you would model accounts as objects with operations for inserting and withdrawing money. You would try to, as much as possible, map concepts from the problem domain to software as directly as possible. This is a powerful approach to designing your software, but there are always aspects of software that do not readily fit into such modeling, especially when it comes to algorithmic programming and design of data structures. Search trees and sorting algorithms, for instance, are usually not reflecting anything concrete in a problem domain.

Object-oriented programming, however, is also a very powerful tool to use when designing algorithms and data structures. The way I was taught programming, algorithms and data structures were covered in separate classes from those in which I was taught object-orientation. Combining object-orientation and algorithmic programming were something I had to teach myself by writing software. I think this was a pity since the two really fit together well.

In this book, I will try to cover object-orientation both as a modeling technique for designing software but also as a tool for developing reusable algorithmic software. Polymorphism, a cornerstone of object-oriented programming, lends itself readily to developing flexible algorithms and to combining different concrete implementations of abstract data types to tailor abstract algorithms to concrete problems. A main use of R is machine learning and data science where efficient and flexible algorithms are more important than modeling a problem domain, so much of the book will focus on those aspects of object-oriented programming.

To read this book, you need to know the fundamentals of R programming: how to manipulate data and how to write functions. We will not see particularly complex R programming, so you do not need a fundamental knowledge of how to do functional programming in R, but should you want to learn how, I suggest reading the first book in this series which is about exactly that. You should be able to follow the book without having read it, though.

CHAPTER 1

■ ■ ■

Classes and Generic Functions

R's approach to object-oriented programming is through *generic functions* and *classes*. As mentioned in the introduction, R actually has three systems for implementing this, called S3, S4, and R6. In this chapter, I will only describe the S3 system, which is the simplest of the three, but I will return to the other two systems in later chapters.

Generic Functions

The term *generic functions* refers to functions that can be used on more than one data type. Since R is dynamically typed, which means that there is no check of type consistency before you run your programs, type checking is really only a question of whether you can manipulate data in the way your functions attempt to. This is also called "duck typing" from the phrase "if it walks like a duck...". If you can do the operations you want to do on a data object, then it has the right type. Where generic functions come into play is when you want to do the same semantic operation on objects of different types, but where the implementation of how that operation is done depends on the concrete types. Generic functions are functions that work differently on different types of objects. They are therefore also known as *polymorphic functions*.

To take this down from the abstract discussion to something more concrete, let us consider an abstract data type, say a stack. A stack is defined by the operations we can do on it, such as the following:

- Get the top element

- Pop the first element off a stack

- Push a new element to the top of the stack

© Thomas Mailund 2017
T. Mailund, *Advanced Object-Oriented Programming in R*, DOI 10.1007/978-1-4842-2919-4_1

To have a base case for stacks we typically also want a way to

- Create an empty stack

- Check if a stack is empty

These five operations define what a stack *is*, but we can implement a stack in many different ways. Defining a stack by the operations that we can do on stacks makes it an abstract data type. To implement a stack, we need a concrete implementation.

In a statically typed programming language, we would define the type of a stack by these operations. How this would be done depends on the type of programming language and the concrete language, but generally in statically typed functional languages you would define a *signature* for a stack—the functions and their type for the five operations—while in an object-oriented language you would define an abstract superclass.

In R, the types are implicitly defined, but for a stack, we would also define the five functions. These functions would be generic and not actually have any implementation in them; the implementation goes into the concrete implementation of stacks.

Of the five functions defining a stack, one is special. Creating an empty stack does not work as a generic function. When we create a stack, we always need a concrete implementation. But the other four can be defined as generic functions. Defining a generic function is done using the UseMethod function, and the four functions can be defined as thus:

```
top <- function(stack) UseMethod("top")
pop <- function(stack) UseMethod("pop")
push <- function(stack, element) UseMethod("push")
is_empty <- function(stack) UseMethod("is_empty")
```

What UseMethod does here is dispatch to different concrete implementations of functions in the S3 object-oriented programming system. When it is called, it will look for an implementation of a function and call it with the parameters that the generic function was called with. We will see how this lookup works shortly.

When defining generic functions, you can specify "default" functions as well. These are called when UseMethod cannot find a concrete implementation. These are mostly useful when it is possible to actually have some default behavior that works in most cases, so not all concrete classes need to implement them. But it is a good idea to always implement them, even if all they do is inform you that an actual implementation wasn't found. For example:

```
top.default <- function(stack) .NotYetImplemented()
pop.default <- function(stack) .NotYetImplemented()
push.default <- function(stack, element) .NotYetImplemented()
is_empty.default <- function(stack) .NotYetImplemented()
```

Classes

To make concrete implementations of abstract data types we need to use *classes*. In the S3 system, you create a class, and assign a class to an object, just by setting an attribute on the object. The name of the class is all that defines it, so there is no real type checking involved. Any object can have an attribute called "class" and any string can be the name of a class.

We can make a concrete implementation of a stack using a vector. To define the class we just need to pick a name for it. We can use `vector_stack`. We create such a stack using a function for creating an empty stack, and in this function, we set the attribute "class" using the `class<-` modification function.

```
empty_vector_stack <- function() {
  stack <- vector("numeric")
  class(stack) <- "vector_stack"
  stack
}

stack <- empty_vector_stack()
stack
## numeric(0)
## attr(,"class")
## [1] "vector_stack"
attributes(stack)
## $class
## [1] "vector_stack"
class(stack)
## [1] "vector_stack"
```

The empty stack is a numeric vector, just because we need some type to give the empty vector, but pushing other values onto it will just force a type conversion, so we can also put other basic types into it. It is limited to basic data types since vectors cannot contain complex data; for that, we would need a list. If we need complex data, we could easily change the implementation to use a list instead of a vector.

We will push elements by putting them at the front of the vector, pop elements by getting everything except the first element of the vector, and of course get the top of a vector by just indexing the first element. Such an implementation can look like this:

```
top.vector_stack <- function(stack) stack[1]
pop.vector_stack <- function(stack) {
  new_stack <- stack[-1]
  class(new_stack) <- "vector_stack"
  new_stack
}
```

3

```
push.vector_stack <- function(element, stack) {
  new_stack <- c(element, stack)
  class(new_stack) <- "vector_stack"
  new_stack
}
is_empty.vector_stack <- function(stack) length(stack) == 0
```

You will notice that the names of the functions are composed of two parts. Before the "." (period) you have the names of the generic functions that define a stack, and after the period you have the class name. This name format has semantic meaning; it is how generic functions figure out which concrete functions should be called based on the data provided to them.

When the generic functions call UseMethod, this function will check if the first value with which the generic function was called has an associated class. If so, it will get the name of that class and see if it can find a function with the name of the generic function (the name parameter given to UseMethod, not necessarily the name of the function that calls UseMethod) before a period and the name of the class after the period. If so, it will call that function. If not, it will look for a .default suffix instead and call that function if it exists.

This lookup mechanism gives semantic meaning to function names, and you really shouldn't use periods in function names unless you want R to interpret the names in this way. The built-in functions in R are not careful about this—R has a long history and is not terribly consistent in how functions are named—but if you don't want to accidentally implement a function that works as a concrete implementation of a generic function, you shouldn't do it.

If we call push on a vector stack, it will, therefore, be push.vector_stack that will be called instead of push.default.

```
stack <- push(stack, 1)
stack <- push(stack, 2)
stack <- push(stack, 3)
stack
## [1] 1 2 3
## attr(,"class")
## [1] "vector_stack"
```

In the push.vector_stack we explicitly set the class of the concatenated new vector. If we didn't do this, we would just be creating a vector—the stack-ness of the second argument to c does not carry on to the concatenated vector—and we wouldn't return a stack. By setting the class of the new vector, we make sure that we return a stack.

The class isn't preserved when we remove the first element of the vector either, which is why we also have to set the class in the pop.vector_stack function explicitly. Otherwise, we would only have a stack the first time we pop an element, and after that, it would just be a plain vector. By explicitly setting the class we make sure that the function returns a stack that we can use with the generic functions again.

```
while (!is_empty(stack)) {
  stack <- pop(stack)
}
```

The remaining two functions, top and is_empty, do not return a stack object, and they are not supposed to, so we don't set the class attribute there.

We can avoid having to set the class attribute explicitly whenever we update it—that is, whenever we return a new value; we never actually modify an object here—by wrapping the class creation code in another function. Such a version could look like this:

```
make_vector_stack <- function(elements) {
  structure(elements, class = "vector_stack")
}
empty_vector_stack <- function() {
  make_vector_stack(vector("numeric"))
}
top.vector_stack <- function(stack) stack[1]
pop.vector_stack <- function(stack) {
  make_vector_stack(stack[-1])
}
push.vector_stack <- function(stack, element) {
  make_vector_stack(c(element, stack))
}
is_empty.vector_stack <- function(stack) length(stack) == 0
```

We are of course still setting the class attribute when we create an updated stack, we are just doing so implicitly by translating a vector into a stack using make_vector_stack. That function uses the structure function to set the class attribute, but otherwise just represent the stack as a vector just like before.

Polymorphism in Action

The point of having generic functions is, of course, that we can have different implementations of the abstract operations. For the stack, we can try a different representation. The vector version has the drawback that each time we return a modified stack we need to create a new vector, which means copying all the elements in the new vector from the old. This makes the operations linear time in the vector size. Using a linked list, we can make them constant time operations. Such an implementation can look like this:

```
make_list_node <- function(head, tail) {
  list(head = head, tail = tail)
}
make_list_stack <- function(elements) {
  structure(list(elements = elements), class = "list_stack")
}
empty_list_stack <- function() make_list_stack(NULL)
top.list_stack <- function(stack) stack$elements$head
pop.list_stack <- function(stack) make_list_
stack(stack$elements$tail)
push.list_stack <- function(stack, element) {
  make_list_stack(make_list_node(element, stack$elements))
}
is_empty.list_stack <- function(stack) is.null(stack$elements)

stack <- empty_list_stack()
stack <- push(stack, 1)
stack <- push(stack, 2)
stack <- push(stack, 3)
stack
## $elements
## $elements$head
## [1] 3
##
## $elements$tail
## $elements$tail$head
## [1] 2
##
## $elements$tail$tail
## $elements$tail$tail$head
## [1] 1
##
## $elements$tail$tail$tail
## NULL
##
##
##
##
## attr(,"class")
## [1] "list_stack"
```

Generally, when working with lists, we would use NULL as the base case to terminate a list. We cannot just wrap a list and use NULL this way when we need to associate a class with the element. You cannot set the class to NULL. So instead we wrap the actual list inside another list where we set the class attribute. The real data is in the elements of this list, but except for having to use this list element of the object, we just work with the list representation as we normally would with linked lists.

We now have two different implementations of the stack interface, but—and this is the whole point of having generic functions—code that uses a stack does not need to know which implementation it is operating on, as long as it only accesses stacks through the generic interface.

We can see this in action in the small function below that reverses a sequence of elements by first pushing them all onto a stack and then popping them off again.

```
stack_reverse <- function(empty, elements) {
  stack <- empty
  for (element in elements) {
    stack <- push(stack, element)
  }
  result <- vector(class(top(stack)), length(elements))
  for (i in seq_along(result)) {
    result[i] <- top(stack)
    stack <- pop(stack)
  }
  result
}

stack_reverse(empty_vector_stack(), 1:5)
## [1] 5 4 3 2 1
stack_reverse(empty_list_stack(), 1:5)
## [1] 5 4 3 2 1
```

Since the stack_reverse function only refers to the concrete stacks through the abstract interface, we can give it either a vector stack or a list stack, and it can do the same operations on both. As long as the concrete data structures all implement the abstract interface correctly then code that only uses the generic functions will be able to work on any implementation.

One single concrete implementation is rarely superior in all cases, so it makes sense that we are able to combine algorithms working on abstract data types with concrete implementations, depending on the particular problem we need to solve. For the two stack implementations they generally work equally well, but as discussed above, the stack implementation has a worst-case quadratic running time while the list implementation has a linear running time. For large stacks, we would thus expect the list implementation to be the best choice, but for small stacks, there is more overhead in manipulating the list implementation the way we do—having to do with looking up variable names and linking lists and such—so for short stacks, the vector implementation is faster.

```
library(microbenchmark)
microbenchmark(stack_reverse(empty_vector_stack(), 1:10),
               stack_reverse(empty_list_stack(), 1:10))
## Unit: microseconds
##                                          expr
##  stack_reverse(empty_vector_stack(), 1:10)
##   stack_reverse(empty_list_stack(), 1:10)
##      min       lq     mean   median       uq
##  195.787 217.2255 245.5698 229.8335 252.4785
##  235.270 257.0180 283.9289 271.5655 289.5300
##       max neval cld
##   965.714   100    a
##  1174.888   100    b
microbenchmark(stack_reverse(empty_vector_stack(), 1:1000),
               stack_reverse(empty_list_stack(), 1:1000))
## Unit: milliseconds
##                                            expr
##  stack_reverse(empty_vector_stack(), 1:1000)
##   stack_reverse(empty_list_stack(), 1:1000)
##       min       lq     mean   median       uq
##  28.22786 32.30359 37.77813 34.25044 36.49355
##  23.38964 24.40046 26.14922 25.47550 26.65488
##        max neval cld
##  124.57810   100    b
##   61.01179   100    a
```

Plotting the time usage for various length of stacks makes it even more evident that, as the stacks get longer, the list implementation gets relatively faster than the vector implementation.

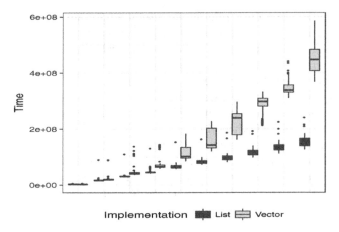

Figure 1-1. *Time usage of reversal with two different stacks*

Only for very short stacks would the vector implementation be preferable—the quadratic versus linear running time kicks in for very small n—but in general, different implementations will be preferred for different usages. By writing code that is polymorphic, we make sure that we can change the implementation of a data structure without having to modify the algorithms using it.

Designing Interfaces

It's not just generic functions that are polymorphic. Any function that manipulates data only through generic functions is also polymorphic. The reversal function we implemented using a stack takes the empty stack as an argument, and this empty stack determines which actual stack implementation we use. Nowhere in the function do we refer to any details of a real implementation. If we had, instead, created an empty stack inside this function then, despite otherwise only accessing the implementation through the interface of the generic functions, the function would be bound to a single implementation.

To get the most out of polymorphism, you will want to design your functions to be as polymorphic as possible. This requires two things:

1. Don't refer to concrete implementations unless you really have to.

2. Any time you *do* have to refer to implementation details of a concrete type, do so through a generic function.

The reversal function is polymorphic because it doesn't refer to any concrete implementation. The choice of which concrete stack to use is determined by a parameter, and the operations it performs on the specific stack implementation all go through generic functions.

It can be very tempting to break these rules in the heat of programming. Using a parameter to determine data structures in an algorithm isn't that difficult to do, but if you are writing an algorithm that uses several different data structures, you might not want to have all the different concrete implementations as parameters. You really ought to do it, though. Just write a function that wraps the algorithm and provides implementations if you don't want to remember all the concrete data structures where the algorithm is needed. That way you get the best of both worlds.

More often, you will want to access the details of a concrete implementation. Imagine, for example, that you want to pop elements until you see a specific one, but *only* if that element is on the stack. If we are used to working with the vector implementation of the stack, then it would be natural to write a function like this:

```
pop_until <- function(stack, element) {
  if (element %in% stack) {
    while (top(stack) != element) stack <- pop(stack)
  }
  stack
}

library(magrittr)
vector_stack <- empty_vector_stack() %>%
  push(1) %>%
  push(2) %>%
  push(3) %T>% print
## [1] 3 2 1
## attr(,"class")
## [1] "vector_stack"
pop_until(vector_stack, 1)
## [1] 1
## attr(,"class")
## [1] "vector_stack"
pop_until(vector_stack, 5)
## [1] 3 2 1
## attr(,"class")
## [1] "vector_stack"
```

Here we use the %in% function to test if the element is on the stack (and we use the magrittr pipe operator to create a stack for our test). This works fine, as long as the stack is a vector stack, but it will *not* work if the stack is implemented as a list. You won't get an error message; the %in% test will just always return FALSE, so if you replace the stack implementation you have incorrect code that doesn't even inform you that it isn't working.

Relying on implementation details is the worst thing you can do to break the interface of polymorphic objects. Not only do you tie yourself to a single implementation, but you also tie yourself to exactly how that concrete data is implemented. If that implementation changes, your algorithm using it will break. So now you either can't change the implementation, or you will have to change the algorithm that it uses when it does. If you are lucky, you might get an error message if you break the interface, but as in the case we just saw (and you can try it yourself if you don't believe me), you won't even get that. The function will just always return the original stack, even when the element you want to pop to is on it.

```
list_stack <- empty_list_stack() %>%
  push(1) %>%
  push(2) %>%
  push(3)
pop_until(list_stack, 1)
```

If you write an algorithm that operates on a polymorphic object, stick to the interface it has, if at all possible. For the pop_until function we can easily implement it using just the stack interface.

```
pop_until <- function(stack, element) {
  s <- stack
  while (!is_empty(s) && top(s) != element) s <- pop(s)
  if (is_empty(s)) stack else s
}
```

If you cannot achieve what you need using the interface, you should instead extend it. You can always write new generic functions that work on a class.

```
contains <- function(stack, element) {
  UseMethod("contains")
}
contains.default <- function(stack, element) {
  .NotYetImplemented()
}
contains.vector_stack <- function(stack, element) {
  element %in% stack
}
```

You do not need to implement concrete functions for all implementations of an abstract data type to add a generic function. If you have a default implementation that gives you an error—and you have proper unit tests for any code you use—you will get an error if your algorithm attempts to use the function if it isn't implemented yet, and you can add it at that point.

Adding new generic functions is not as ideal as using the original interface in the first place if the abstract data type is from another package. If the implementation in that package changes at a later point, your new generic function might break—and might break silently. Still, combined with proper unit tests, it is a much better solution than simply accessing the detailed implementation in your other functions.

Designing interfaces is a bit of an art. When you create your own abstract types, you want to think carefully about which operations the type should have. You don't want to have too many operations. That would make it harder for people implementing other versions of the type; they would need to implement all the operations, and depending on what those operations are, this could involve a lot of work. On the other hand, you can't have too few operations, because then algorithms using the type will often have to break the interface to get to implementation details, which will break the polymorphism of those algorithms.

The abstract data types you learn about in an algorithms class are good examples of minimal yet powerful interfaces. They define the minimum number of operations necessary to get useful work done, yet still make implementations of concrete stacks, queues, dictionaries, etc. possible with minimal work.

When designing your own types, try to achieve the same kind of minimal interfaces.

The Usefulness of Polymorphism

Polymorphism isn't only useful for what we would traditionally call abstract data structures. Polymorphism gives you the means to implement abstract data structures, so algorithms work on the abstract interface and never need to know which concrete implementation they are operating on, but generic functions are useful for many cases that we do not traditionally think of as data structures.

In R, you often fit statistical models to data. Such models are not really data structures, but there is an abstract interface to them. You fit a concrete model, for example, a linear model, but once you have a fitted model, there are many common operations that are useful for all models. You might want to predict response variables for new data, or you might want to get the residuals of your fitted values. These operations are the same for all models—although how different models implement the operations will be different—and so they can benefit from being generic. Indeed, they are. The functions predict and residuals, which implement those two operations, are generic functions, and each model can implement its own version of them.

There is an extensive list of standard functions that are frequently used on fitted models, and all of these are implemented as generics. If you write analysis code that operates on fitted models using only those generic functions, you can change the model at any time and reuse all the code without modifying it.

The same goes for printing and plotting functions. Both `print` and `plot` are generic functions, and they have concrete implementations for different data types (and usually also for different fitted models). It is not something we think much about from day to day, but if we didn't have generic functions like these, we would need to use different functions for displaying vectors and for displaying matrices, for example.

Converting between different data types is also a frequent operation, and again polymorphism is highly useful (and frequently used in R). To translate a data structure into a vector, you use the `as.vector` function—an unfortunate name since it looks like a generic function `as` with a specialization for `vector`, but actually is a generic function named `as.vector`. To translate a factor into a vector, it is the concrete implementation `as.vector.factor` that gets called.

An algorithm that needs to translate some input data into a vector can use the `as.vector` function and then doesn't have to worry about what the actual data is implemented as, as long as the data type has an implementation of the `as.vector` function.

Polymorphism and Algorithmic Programming

Polymorphism as a component of designing algorithms, and especially implementing algorithms, is not often covered in classes and textbooks but can be an important aspect of writing reusable software. You might not think of R as a language where you implement algorithms, but whenever you write a data analysis pipeline, whenever you manipulate data frames, and whenever you fit a model, you can think of that as implementing or using an algorithm. We want our data analysis to be efficient, so we want our algorithms to be efficient, but we also want to write code that can be used more than once so we don't have to repeat ourselves. This means that we need to write code that can be used with different data and in many instances this involves hiding concrete data behind generic interfaces.

Take something as simple as a sorting function. For many sorting algorithms, all you need to be able to do to sort elements is determine whether one element is smaller than another. If you hardwire in an implementation of such an algorithm where the comparison used is interfering or floating point comparison, then you can only sort objects of these types. In general, if you hardwire comparisons, you need a different implementation for each type of elements you want to sort.

Because of this, most languages provide you with a generic sorting function as part of their runtime library where you can provide the comparison functionality it should use, typically either as a function provided to the function or by allowing you to specify a comparison function for new types. Unfortunately, the `sort` function in R is not of this kind—it does allow you to define sorting for new types, but it wants

13

its input to be in atomic form, so you cannot give it sequences of complex data types—anything beyond simple numerical, boolean, or string types. Usually, you can change your data to a matrix or something similar and sort it this way, but if you actually have a list of complex data, you cannot use it.

We can easily implement our own function for doing this, however, and we can call it sort_list—not to be confused with the built-in function sort.list that actually does something other than sort lists...

Sorting Lists

A straightforward implementation of merge sort could look like this:

```
merge_lists <- function(x, y) {
  if (length(x) == 0) return(y)
  if (length(y) == 0) return(x)

  if (x[[1]] < y[[1]]) {
    c(x[1], merge_lists(x[-1], y))
  } else {
    c(y[1], merge_lists(x, y[-1]))
  }
}

sort_list <- function(x) {
  if (length(x) <= 1) return(x)

  start <- 1
  end <- length(x)
  middle <- end %/% 2

  merge_lists(sort_list(x[start:middle]),
              sort_list(x[(middle+1):end]))
}
```

It gets the job done, but the merge function is quadratic in running time since it copies lists when it subscripts like x[-1] and y[-1] and when it combines the results in the recursive calls. We can make a slightly more complicated function that does the merging in linear time using an iterative approach rather than a recursive:

```
merge_lists <- function(x, y) {
  if (length(x) == 0) return(y)
  if (length(y) == 0) return(x)
```

```
  i <- j <- k <- 1
  n <- length(x) + length(y)
  result <- vector("list", length = n)

  while (i <= length(x) && j <= length(y)) {
    if (x[[i]] < y[[j]]) {
      result[[k]] <- x[[i]]
      i <- i + 1
    } else {
      result[[k]] <- y[[j]]
      j <- j + 1
    }
    k <- k + 1
  }

  if (i > length(x)) {
    result[k:n] <- y[j:length(y)]
  } else {
    result[k:n] <- x[i:length(x)]
  }

  result
}
```

We are still copying in the recursive calls of the sorting function, but we are not copying more than we will merge later, so the asymptotic running time is okay, at least.

With this function, we can sort lists of elements where "<" can be used to determine if one element is less than another. The built-in "<" function, however, doesn't necessarily work on your own classes.

```
make_tuple <- function(x, y) {
  result <- c(x,y)
  class(result) <- "tuple"
  result
}

x <- list(make_tuple(1,2),
          make_tuple(1,1),
          make_tuple(2,0))
sort_list(x)
## Warning in if (x[[i]] < y[[j]]) {: the condition
## has length > 1 and only the first element will be
## used
```

15

```
## Warning in if (x[[i]] < y[[j]]) {: the condition
## has length > 1 and only the first element will be
## used

## Warning in if (x[[i]] < y[[j]]) {: the condition
## has length > 1 and only the first element will be
## used
## [[1]]
## [1] 1 1
## attr(,"class")
## [1] "tuple"
##
## [[2]]
## [1] 1 2
## attr(,"class")
## [1] "tuple"
##
## [[3]]
## [1] 2 0
## attr(,"class")
## [1] "tuple"
```

There are several ways we can fix this. One option is to define a generic function for comparison: we could call it less, and then use that in the merge function.

```
merge_lists <- function(x, y) {
  if (length(x) == 0) return(y)
  if (length(y) == 0) return(x)

  i <- j <- k <- 1
  n <- length(x) + length(y)
  result <- vector("list", length = n)

  while (i <= length(x) && j <= length(y)) {
    if (less(x[[i]], y[[j]])) {
      result[[k]] <- x[[i]]
      i <- i + 1
    } else {
      result[[k]] <- y[[j]]
      j <- j + 1
    }
    k <- k + 1
  }
```

```
  if (i > length(x)) {
    result[k:n] <- y[j:length(y)]
  } else {
    result[k:n] <- x[i:length(x)]
  }

  result
}

less <- function(x, y) UseMethod("less")
less.numeric <- function(x, y) x < y
less.tuple <- function(x, y) x[1] < y[1] || x[2] < y[2]

sort_list(x)
## [[1]]
## [1] 1 1
## attr(,"class")
## [1] "tuple"
##
## [[2]]
## [1] 1 2
## attr(,"class")
## [1] "tuple"
##
## [[3]]
## [1] 2 0
## attr(,"class")
## [1] "tuple"
```

We would need to define concrete implementations of less for all types we wish to sort, though. Alternatively, we can tell R how to handle "<" for our own types, and we will see how in a later chapter. With that approach, we will get sorting functionality for all objects that can be compared this way. A third possibility is to make less a parameter of the sorting function:

```
merge_lists <- function(x, y, less) {
  # Same function body as before
}
sort_list <- function(x, less = `<`) {

  if (length(x) <= 1) return(x)

  result <- vector("list", length = length(x))
```

17

```
  start <- 1
  end <- length(x)
  middle <- end %/% 2

  merge_lists(sort_list(x[start:middle], less),
              sort_list(x[(middle+1):end], less),
              less)
}

unlist(sort_list(as.list(sample(1:5))))
## [1] 1 2 3 4 5
tuple_less <- function(x, y) x[1] < y[1] || x[2] < y[2]
sort_list(x, tuple_less)
## [[1]]
## [1] 1 1
## attr(,"class")
## [1] "tuple"
##
## [[2]]
## [1] 1 2
## attr(,"class")
## [1] "tuple"
##
## [[3]]
## [1] 2 0
## attr(,"class")
## [1] "tuple"
```

We make the default less function "<" but can provide another for types where this comparison function doesn't work.

General Comments on Flexible Implementations of Algorithms

As a general rule, you want to make your algorithm implementations adaptable by providing handles for polymorphism, either by providing options for certain functions (like we did with less above) or by using generic functions for abstract data types.

You might be able to experiment with optimal data structures and implementation of operations when you implement an algorithm for a given use, but by providing handles for modifying your function you make the code more reusable. Even in cases where the algorithm will perform correctly for different applications, you might still want to provide flexibility; the performance of algorithms often depends on the usage. In an asymptotic analysis we generally prefer implementations that have theoretical better running times, but in practice, we want the fastest code, and that is not necessarily the asymptotically fastest algorithms. We hide away constants when we use "big-O" analysis, but those constants matter, so you want users of your implementations to be able to replace data structures and operations used in your algorithm implementations.

Figuring out how to best provide this flexibility in your implementations often requires some experimentation. For abstract data structures, generic functions are usually the best approach. For something like comparison in the sorting example above, all three solutions (generic functions, operator overloading, or providing a function with a good default) are probably equally good. But just as experimentation and some thinking are involved in designing good software interfaces, the same is needed in algorithmic programming.

More on UseMethod

The UseMethod function is what we use to define a generic function, and it takes care of finding the appropriate concrete implementation using the name lookup we saw earlier. There are some details about UseMethod I left out before, though.

First of all, it doesn't actually work as a function normally does. It looks like a function, and to a large degree it is a function, but if you treat it just as any other function you might get effects you didn't expect.

Second of all, you can pass local variables along to concrete implementations if you assign them before you call UseMethod. Let's consider a simple case.

```
foo <- function(object) UseMethod("foo")
foo.numeric <- function(object) object
foo(4)
## [1] 4
bar <- function(object) {
  x <- 2
  UseMethod("bar")
}
bar.numeric <- function(object) x + object
bar(4)
## [1] 6
```

Here the foo function uses the pattern we saw earlier. It just calls UseMethod. We then define a concrete function to be called if foo is invoked on a number. Numbers have classes, and that class is numeric. (Technically, there is more to numbers than this class, but for now, we don't need to worry about that.) Nothing strange is going on with foo.

With bar, however, we assign a local variable before we invoke UseMethod. This variable, x, is visible when bar.numeric is called. With a normal function call, you have to take steps to get access to the calling scope, so here UseMethod does not behave like a normal function.

In the call to UseMethod, it doesn't behave like a normal function either. You cannot use UseMethod as part of an expression.

```
baz <- function(object) UseMethod("baz") + 2
baz.numeric <- function(object) object
baz(4)
## [1] 4
```

When UseMethod is invoked, the concrete function takes over completely, and the call to UseMethod never returns. In this way, it is similar to the return function. Any expression you put UseMethod in is not evaluated because of this, and any code you might put after the UseMethod call is never evaluated.

The UseMethod function takes a second argument, besides the name of the generic function. This is the object that is used to dispatch the generic function on—the object whose type determines the concrete function that will be called—and this argument can be used if you do not want to dispatch based on the first argument of the function that calls UseMethod. Since dispatching on the type of the first function argument is such a common pattern, using another object in the call to UseMethod can cause confusion, and I recommend that you do not do this unless you have very good reasons for it.

CHAPTER 2

■ ■ ■

Class Hierarchies

There is more to polymorphism than merely abstract data types that can have different concrete implementations. A fundamental concept found in most object-oriented programming languages is classes and class hierarchies. Class hierarchies serve two conceptually different purposes: refinement of object interfaces and code reuse. Neither concept, strictly speaking, requires class hierarchies in R, since R is dynamically typed, unlike programming languages such as C++ or Java where class hierarchies and the static type system are intertwined. Nevertheless, class hierarchies provide a framework for thinking about software design that is immensely useful, not least in dynamically typed languages.

We will go into details of the two concepts in the two following sections, but in short, interfaces describe which (generic) functions objects of a given class must implement, and hierarchies chain together interfaces in "more-abstract/more-refined" relationships based on these functions. Code-reuse, in this context, refers simply to writing functions that can operate on more than one class of objects—essentially just the type of polymorphic functions we saw in the previous chapter—and fitting such functions into class hierarchies as generic functions themselves.

Interfaces and Implementations

We can think of the *interface* of a class as the kinds of operations, or methods, which we can apply to objects of the class. In R, this means which functions we can call with such objects as arguments in a meaningful way.

If we think in terms of abstract data structures, such as the stack from the last chapter, these are defined by which operations they support. You can *push* and *pop* from a stack, check if it is *empty*, and you can get the *top* element; those functions, together with a way of creating a stack, define what "stack-ness" *is*. At least, as long as those four functions also have the semantics we associate with a stack.

© Thomas Mailund 2017
T. Mailund, *Advanced Object-Oriented Programming in R*, DOI 10.1007/978-1-4842-2919-4_2

At an abstract level, we can describe the interface of a function by its formal arguments and its semantics. We can, for example, associate with a function *push* its two formal parameters, a stack and an element, and the semantics that it should return the stack but with the element added to the top. If we associate the *push* operation with these two attributes, the formal parameters and the semantics, we have what we could call an *abstract function*. As we saw, we can implement such abstract functions in different concrete ways, but a caller of these concrete functions need only worry about the abstract description to ensure correctness of functionality (although performance can of course also be a concern and not something we associated with the interface of an abstract function here).

With this definition of abstract functions, we can say that an abstract data type is defined by a set of abstract functions. If we call a set of abstract functions an *interface*, then an abstract data type is defined by an interface. We can implement an abstract data structure by writing an implementation of all the abstract functions in the interface. This we might call a (concrete) *implementation* of the interface or something along those lines. We can reason about algorithms and design software just from knowing the interface of an abstract data type, and if we have different implementations of the interface to choose from, then any of them could, in theory, be used.

Concepts such as interfaces and implementations are not just useful when it comes to abstract data structures. For any type of data you want to manipulate in a program, you could think up a set of meaningful operations you could do on that data, thus creating an interface for the type of data, and you could write functions for those operations in different ways to create different implementations.

Polymorphism and Interfaces

If we go back to thinking about interfaces, we can say that a class implements an interface if it implements all the abstract functions that make up the interface. This simply means that, if we take objects of this class, we have concrete functions we can call for each of the abstract functions in the interface. Without generic/polymorphic functions, however, we would need to know which concrete function maps to which abstract function for each class that implements a given interface. Exchanging one implementation of an interface with another would require a rewrite of the code that uses the implementation. So naturally, we would also require that the names of the concrete functions match the names of the abstract functions.

This obviously maps directly to generic functions. If, whenever we think *abstract function*, we map that to a *generic function*—one that simply calls UseMethod—and whenever we think *concrete function* we think implementation of a generic function—a function with a period in its name—then we have an almost automatic way of mapping the concepts of interfaces and implementations into code.

Since R doesn't do any static type checking, there is very little you can do to guarantee that a class you write this way actually implements a given interface. There is nothing in generic functions that explicitly binds them together as an interface, so for any class you decide to implement, you can implement an arbitrary subset of generic functions. Interfaces and implementations are design concepts, and you can map the design into R code very easily, but R does not enforce that your code matches your design.

Abstract and Concrete Classes

We often unify interfaces and implementations as just classes, at least when designing software. The object-oriented way to think about software is this: every piece of data you manipulate is an *object* and all objects have a *class* that determines their behavior. By "behavior," we just mean which functions we can call on an object. This way of thinking makes a little more sense in languages where you can modify data and where objects thus have a state. Regardless, you can think of all data as objects with associated classes that determine what you can do with them.

A *class* thus encapsulates both what you can do with objects—the interface you have for them—but also how it is done—how the interface is implemented.

Objects have classes, and classes determine what you can do with objects, but classes live in hierarchies of more abstract or more derived classes. A vector-based implementation of a stack is a stack. It is a special *kind* of stack, sure, but it is still a stack. The general concept of what a stack *is* is more general than vector-based implementations, so the vector implementation can be thought of as a specialisation of a stack—that is, one that is implemented using a vector.

We generally think about class hierarchies as part of "is-a" relationships. A vector implementation of a stack "is-a" stack. So is a list-based implementation. If you have an object of a more specialised class you should also be able to treat it as an object of a more abstract class. If you have a vector stack, you can treat it as a stack because its class is a vector stack class and that is a special case of the stack class.

The closest we get to interfaces and implementations is *abstract classes* and *concrete classes*. An abstract class is essentially exactly an interface. It is nothing more than a description of what you can do with objects of this class; there is no implementation associated with it. Concrete classes, on the other hand, have implementations for all the functions you can call on objects of the given class. Quite often, though, classes implement some but not all the functions their interface describes, so the distinction is not that clean in practise.

We often show classes and their relationships in diagrams as that shown in Figure 2-1. Here *Stack* is shown in italics to indicate that it is an abstract class. Below the class name is listed the methods you can call on the class, and errors from one class to another indicate that one class is derived from another. Here we see that vector and list stacks, here called *VectorStack* and *ListStack* are derived from *Stack*.

23

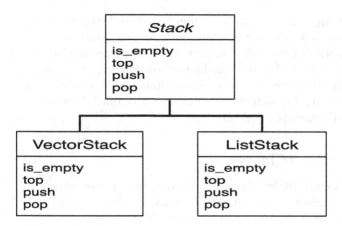

Figure 2-1. *Class hierarchy for stacks*

The two concrete classes only implement the methods also listed in the abstract class, and because of this, we won't always list the methods again in the derived classes. It is to be understood that any method implemented in a more abstract class will also be implemented in more derived classes.

Implementing Abstract and Concrete Classes in R

We already saw, in the previous chapter, how the attribute `class` is used to determine which version of a generic function is called for a given object. This approach for dispatching generic functions is the S3 system's way of implementing classes, but in some sense only handles concrete implementations of abstract functions. Having a generic method `foo`

```
foo <- function(object) UseMethod("foo")
```

that we implement for a class `bar`

```
foo.bar <- function(object) ...
```

only tells R how class `bar` implements the `foo` function. If `foo` is part of an interface that consists of several functions, it is not explicitly stated in the R code.

If we think of interfaces as a set of abstract functions, then considering these as part of a whole is something we only do informally in R. Since abstract classes are nothing more than interfaces, we can do the same for abstract classes. When we implemented the vector-based stack in the previous chapter, we did so by setting the `class` attribute of the objects we returned from the constructor function `empty_vector_stack` to `vector_stack` and by implementing the four functions we considered part of the stack interface: push, pop, top, and is_ empty. At no point did we specify that there existed some abstract stack class and that vector_stack is a specialization of it.

Since the class mechanism implemented this way is essentially working on a per-function level—we have generic functions and implementations of these that are dispatched based on their name—classes and their relationships can be a very messy affair in R. You can alleviate this by thinking about your software design in a more structured way than the language requires. Design your software with classes in mind, implement abstract classes by defining a set of generic functions—you can use comments to group them together and to document that these constitute an interface. Make sure that when you define a concrete class implementing an interface that you don't forget about any of the functions in the interface. You might not implement them all; sometimes there are good reasons to, and sometimes you are just being pragmatic and not implementing something that might be difficult to achieve, but that you don't need yet. Make sure that this is a conscious choice, though, and that you haven't simply forgotten a function.

You can use the function methods to get a list of all the methods implemented by a class

```
methods(class = "vector_stack")
## [1] is_empty pop      push     top
## see '?methods' for accessing help and source code
```

and check if you have everything implemented. You can also use this function to get a list of all classes that implement a given generic function.

```
methods("top")
## [1] top.default      top.list_stack
## [3] top.vector_stack
## see '?methods' for accessing help and source code
```

Another Example: Graphical Objects

The "is-a" relationship underlying a class hierarchy is more flexible than just having abstract classes and implementations of these. It provides us with both a way of modeling that some objects really are of different but related classes, and it provides us with a mechanism for thinking about interfaces as specializations of other interfaces.

Let us consider, for example, an application where we operate on some graphical object—perhaps as part of a new visualization package. The most basic class of this application is the *GraphicalObject* whose objects you can draw. Being able to draw objects is the most basic operation we need for graphical objects. Graphical objects also have a "bounding box"—a rectangle that tells us how large the shape is, something we might need when drawing objects.

This class is abstract, not just because we are defining an interface so we can have different implementations, like with did with the stack, but because it doesn't really make sense to *have* a graphical interface at this abstract level. A concrete class that it does make sense to have objects of is *Point*, which is a graphical object representing a single point. Other classes could be *Circle* and *Rectangle*.

For dealing with more than one graphical object, in an interface which makes that easy, we also have a class, *Composite*, that captures a collection of graphical objects.

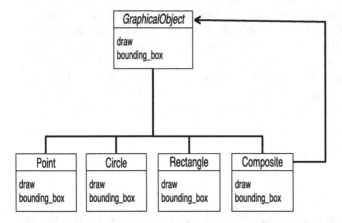

Figure 2-2. *Class hierarchy for graphical objects. The arrow from Composite to GraphicalObject, with a diamond starting point and an arrow endpoint, indicates that a Composite consists of a collection of GraphicalObjects.*

Treating a collection of objects as an object of the same class as its components is a so-called *design pattern* and it makes it easier to deal with complex figures in this application. We can group together graphical objects in a hierarchy—similar to how you would group objects in a drawing tool—and we would not need to explicitly check in our code if we are working on a single object or a collection of objects. A collection of objects is also a graphical object, and we can just treat it as such.

Implementing this class hierarchy is fairly straightforward. The abstract class *GraphicalObject* is not explicitly represented, but we need its methods as generic functions.

```
draw <- function(object) UseMethod("draw")
bounding_box <- function(object) UseMethod("bounding_box")
```

When constructing graphical objects, we need to set their class, and these could be the constructors for the concrete classes:

```
point <- function(x, y) {
  object <- c(x, y)
  class(object) <- "point"
  names(object) <- c("x", "y")
  object
}

rectangle <- function(x1, y1, x2, y2) {
  object <- c(x1, y1, x2, y2)
  class(object) <- "rectangle"
  names(object) <- c("x1", "y1", "x2", "y2")
  object
}

circle <- function(x, y, r) {
  object <- c(x, y, r)
  class(object) <- "circle"
  names(object) <- c("x", "y", "r")
  object
}

composite <- function(...) {
  object <- list(...)
  class(object) <- "composite"
  object
}
```

We just implement the graphical objects as vectors, except for the composite that we represent as a list so it can contain different types of other graphical objects. The points are just vectors of coordinates, the rectangles are represented by two coordinates, the rectangle's lower left and upper right corners, and circles are represented by a center point and a radius.

For the draw methods, we can just use basic graphics functions:

```
draw.point <- function(object) {
  points(object["x"], object["y"])
}

draw.rectangle <- function(object) {
  rect(object["x1"], object["y1"], object["x2"], object["y2"])
}
```

```
draw.circle <- function(object) {
  plotrix::draw.circle(object["x"], object["y"], object["r"])
}

draw.composite <- function(object) {
  invisible(Map(draw, object))
}
```

We can use basic graphics functions except for the circles, where we use the draw.circle function from the plotrix package for convenience. For the collection class, we just call draw on all of the collection's components. We wrap the call to Map in invisible because we don't want to print a list of objects every time we call the function on the command prompt, but otherwise the implementation is straightforward.

With these functions, we can construct plots of graphical elements, see Figure 2-3.

```
plot(c(0, 10), c(0, 10),
     type = 'n', axes = FALSE, xlab = '',*ylab = '')
draw(point(5,5))
draw(rectangle(2.5, 2.5, 7.5, 7.5))
draw(circle(5, 5, 4))

corners <- composite(point(2.5, 2.5), point(2.5, 7.5),
                     point(7.5, 2.5), point(7.5, 7.5))

draw(corners)
```

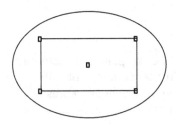

Figure 2-3. *Plot of graphical elements*

Here we have to set the size of the plot so it actually contains the elements we want to display. We have the bounding_box function for calculating what that area is, and we can implement the different methods like this:

```
bounding_box.point <- function(object) {
  c(object["x"], object["y"], object["x"], object["y"])
}
```

```
bounding_box.rectangle <- function(object) {
  c(object["x1"], object["y1"], object["x2"], object["y2"])
}

bounding_box.circle <- function(object) {
  c(object["x"] - object["r"], object["y"] - object["r"],
    object["x"] + object["r"], object["y"] + object["r"])
}

bounding_box.composite <- function(object) {
  if (length(object) == 0) return(c(NA, NA, NA, NA))

  bb <- bounding_box(object[[1]])
  x1 <- bb[1]
  y1 <- bb[2]
  x2 <- bb[3]
  y2 <- bb[4]

  for (element in object) {
    bb <- bounding_box(element)
    x1 <- min(x1, bb[1])
    y1 <- min(y1, bb[2])
    x2 <- max(x2, bb[3])
    y2 <- max(y2, bb[4])
  }

  c(x1, y1, x2, y2)
}
```

With that, we can collect all the graphical elements we wish to plot in a composite object and calculate the bounding box before we plot.

```
all <- composite(
  point(5,5),
  rectangle(2.5, 2.5, 7.5, 7.5),
  circle(5, 5, 4),
  composite(point(2.5, 2.5), point(2.5, 7.5),
            point(7.5, 2.5), point(7.5, 7.5))
)
bb <- bounding_box(all)
plot(c(bb[1], bb[3]), c(bb[2], bb[4]),
     type = 'n', axes = FALSE, xlab = '', ylab = '')
draw(all)
```

The result is, of course, just the same as seen in Figure 2-3.

Class Hierarchies As Interfaces with Refinements

In the examples so far, we have had an abstract class defining an interface and then different concrete classes implementing it. In the case of the stack, the different implementations gave us different time-complexity tradeoffs, but the different implementations were conceptually all just stacks. In the case of the graphical objects, the various concrete classes were conceptually different objects, just objects that can all be treated as graphical objects and thus manipulated through the general interface. These are common patterns in software design, but when sub-classes (such as the different types of graphical objects) represent different conceptual classes, they can often also extend the interface.

Take, for instance, statistical models. These are usually implemented as classes that implement some generic functions, such as predict or coef, that give us a uniform interface to models and make it possible to switch between different models in analysis without major rewrites of our analysis code. The generic functions implemented by all models give us an interface for the most abstract kind of models. All models must implement predict and coef, for example, for us to be able to use them as drop-in replacements in our analysis code, but different types of models might add additional functionality to this interface that is not relevant for all models. We could, for example, imagine that decision trees add functionality for pruning trees, e.g. a function prune. If all decision tree implementations have a prune function, we can replace the implementation of decision trees and still reuse our code, but because prune is not implemented for all models, we can only replace one implementation of a decision tree with another decision tree, not any kind of model. We would say that decision trees are specializations of models. All decision trees are models, but not all models are decision trees. In term of classes, we would have a super-class for models and a sub-class for decision trees that adds to the interface of models functions such as prune. If we have different implementations of decision trees, the decision tree class would typically also be abstract, and different implementations would inherit from this class rather than the more general model class (see Figure 2-4).

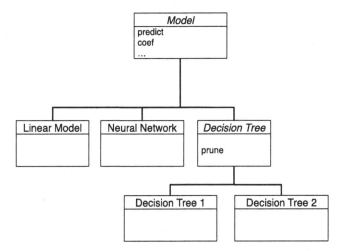

Figure 2-4. *Hierarchy of models where decision trees are specializations of the Model class that adds the* prune *function and gives us an additional abstract class that instances of decision trees must implement*

As another example, we can consider a bibliography, which is essentially a list of publications. There are different kinds of publications, but all have at least a name and one or more authors, so the most abstract way of representing publications would just have those two attributes. One thing we might want to do with a list of publications is to calculate bibliometrics, such as how many citations a publication has. If each publication has a list of other works it cites, then we could calculate this from a database of all relevant publications. If we have a list of all publications a given author has created, we could also calculate how many citations this particular author has received or derive statistics such as the h-index.

There are different types of publications, so we can create sub-classes for books and articles, for example. A book will have an associated publisher and an ISBN while journal articles will have an associated journal and (typically) the page-numbers in the journal where the article can be found. If we, for simplicity, only allow those two types of publications, we can represent it as the class hierarchy shown in Figure 2-5.

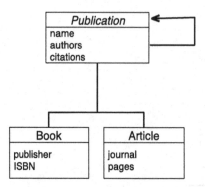

Figure 2-5. *Hierarchy of bibliography objects. The generic Publication class gives each publication a name and a list of authors and a list of other publications cited. Two concrete types of publications, Article and Book, add extra attributes.*

In this hierarchy, all the functions simply access attributes of objects; that is, they just extract data that is stored in these. Accessing attributes via functions, as opposed to accessing lists or vectors that are part of an object's representation, is generally a good idea. It allows you to change how you represent data without having to change any code besides the accessor function. If your code only accesses your objects through functions, then you have encapsulated the implementation details, and updating your code later will be much simpler than it would otherwise be.

Notice also that none of the accessor functions needs to differ between the two concrete types of publications. The abstract *Publication* class accesses name and authors and the additional attributes provided in the other classes are disjunct. Because of this, there is no need to have generic functions for implementing this class hierarchy. We can do it using just plain old functions.

```
# publication interface
publication <- function(name, authors, citations) {
  structure(list(name = name, authors = authors,
                 citations = citations),
            class = "publication")
}
name <- function(pub) pub$name
authors <- function(pub) pub$authors

# articles
article <- function(name, authors, citations, journal, pages) {
  structure(list(name = name, authors = authors,
                 citations = citations,
                 journal = journal, pages = pages),
            class = c("article", "publication"))
}
```

```
journal <- function(pub) pub$journal
pages <- function(pub) pub$pages

# book
book <- function(name, authors, citations, publisher, ISBN) {
  structure(list(name = name, authors = authors,
                 citations = citations,
                 publisher = publisher, ISBN = ISBN),
          class = c("book", "publication"))
}
publisher <- function(pub) pub$publisher
ISBN <- function(pub) pub$ISBN
```

Generic functions are perfect for getting different behavior in different classes for the same conceptual operation, but when we can get the action we desire using plain old functions there is no reason to invoke generic functions.

The implementation of publications is straightforward except when it comes to the class attributes set in the constructor functions. Here we set the classes to lists of class names instead of just the class names. This is how we specify that a "book" object or an "article" object is also a "publication" object. The design we have in mind requires that books and articles are publications, but since S3 classes are just names represented as strings, we cannot make this explicit in R. Instead we represent the class hierarchy by having the class attributes be lists of class names, going up the hierarchy from the most specialized to the most abstract object. How R interprets such a list of class names, and how it uses it to find the right implementations of generic functions, is the topic of the next chapter.

It is not uncommon to have a class hierarchy similar to the one we made here for publications, but there are some slight problems with it. To access book-specific attributes, you need to know that the object you are working on is a book; treating publications in aggregates without having to write specialized code for dealing with books or articles is the purpose of using object-orientation and having the publication hierarchy to begin with. There is nothing wrong with having a class hierarchy where sub-classes add functions to the interface of their superclass, but if you find yourself writing such a hierarchy, you should think carefully about how objects from the hierarchy should be accessed and manipulated.

It is generally best to put functions as high up in the hierarchy as it makes sense to do, thus ensuring that as many classes from the hierarchy as possible will support them. With generic functions, the different subclasses can implement the methods very differently, but all objects you manipulate will at least implement the methods you call them on. With the publication class hierarchy we have designed here, the only things we can really do if we want to write reusable code are to access names and authors and construct graphs of citations. The special attributes for books and articles are only available in a type-safe way if we explicitly check that we are accessing books and articles, respectively.

We would probably like code to format publications for making publication lists and such. Here we would need the information stored in books and articles, but since accessing these directly requires that we first test the class of the objects, the code would be a bit cumbersome and likely also error-prone. Worse, if we added another publication type to the hierarchy (for example, conference contributions), we would need to update all code that does this class checking and handles the different classes in different ways to handle this type as well. It is exactly to avoid this that we need generic functions, so the right design would be to have a generic function for formatting citations and specialize it for the sub-classes.

```
format <- function(publ) UseMethod("format")

format.article <- function(publ) {
  paste(name(publ), authors(publ),
        journal(publ), pages(publ), sep = ", ")
}

format.book <- function(publ) {
  paste(name(publ), authors(publ),
        publisher(publ), ISBN(publ), sep = ", ")
}
```

We could then use this `format` function in another generic function, `print`, for displaying publications:

```
print.publication <- function(x, ...) print(format(x))
```

When we subclass to extend an interface, we add functions that only a subset of objects will support. Sometimes this is necessary when there are operations that truly only make sense for some objects—like pruning decision trees, where pruning something like a linear model is not meaningful—but as a general rule, I would suggest that you keep specialization like this to a minimum. It might feel like a good design to have a large hierarchy of more or less specialized classes, but when you have to work with objects from the hierarchy you want them to be as similar as you can get them. You want to be able to treat all of them using the same (generic) functions, so, in general, you want to stick to the most abstract interface in any case. You might as well design your code with that in mind.

Being able to treat objects uniformly is also the reason we made a collection of graphical objects be a graphical object in itself. If we had not, then we would need to explicitly deal with collections of objects and write recursive functions to traverse them. By making the collection a class of graphical objects, we could hide this complexity in the generic functions. This is a very common trick and is called the composite design pattern.

CHAPTER 3

Implementation Reuse

The easiest code to write is the code that is already written. If you can reuse
existing code, it is always better than writing new code. You don't have to spend
time writing the code, and if the existing code has been around for a while there
is a good chance that it is well tested and thus more likely to be correct than the
new code you introduce. So your aim as a programmer should be to write as
little new code as possible, paradoxical as it sounds.

The way to reuse code, however, is not to copy chunks of existing code from
one place in your program and paste it into another. You will be reusing the code
this way, true, but you will end up with two copies of the same code. If you find, at
some point in the future, that you have to modify this code, perhaps because there
was a bug hidden in it despite the tests, you will have to remember to change it
both places. It is hard to remember when you have such duplicated code and if the
two copies start drifting apart, it is a nightmare trying to maintain it.

The way to reuse code is to write functions that can be used in many
different contexts. Write functions to deal with a single problem, and deal with
it well, while assuming little about the context in which they will be called.
Constructing a program out of small, reusable building blocks makes your
code easier to maintain and simpler to extend.

In object-oriented programming, we often add another aspect to code
reuse, however. When a class defines generic functions, all implementations
of the class and its subclasses must implement these generic functions. If
each class implements all of the generic functions in unique ways, there is
little opportunity for reuse, but it is rarely the case that *everything* is unique.
Often there are good default solutions to a given problem that will work for
most classes, while only a few classes need to implement special functions. In
object-oriented programming, you want to implement functions at the highest
level in the class hierarchy where it is possible and let subclasses reuse these
implementations when possible by calling general functions instead of providing
their own specialized functions. Any class in the hierarchy might have to
specialize a subset of the generic functions, and any subclass of should use the
most specialized function in the hierarchy above it, but in general it should only
define its own version of generic functions if they are different from the ones

© Thomas Mailund 2017

T. Mailund, *Advanced Object-Oriented Programming in R*, DOI 10.1007/978-1-4842-2919-4_3

defined in its superclasses. The remaining functions it should just reuse. We say that it *inherits* such functions from its superclass and the look-up mechanism for generic functions automates this to a high degree.

Deep class hierarchies are rarely used in day-to-day R programming, and rather than making up an artificial example that wouldn't be realistic for actual code—and rather than presenting examples where much of the code would not be relevant for the presentation of class hierarchies—I will use abstract examples. If you are more interested in realistic use of object-orientated programming in R, I give a realistic example in Chapter 4, although one that is only a single class deep.

Method Lookup in Class Hierarchies

As we saw earlier, for a generic function f calling f(x) will make R look for a function with a name derived from class(x). If class(x) is a single string, say "A", it will look for f.A. If R finds such a function it will call it, and if it doesn't, it will look for f.default and call that. Otherwise, it will give up and give you an error if there is no f.default. This is how R behaves if the class of an object is a single string. If it is a sequence of class names, it instead goes through this sequence and attempts to find functions for each of the classes in the sequence in turn.

Consider the three classes A, B, and C, defined below.

```
A <- function() {
  structure(list(), class = "A")
}

B <- function() {
  structure(list(), class = c("B", "A"))
}

C <- function() {
  structure(list(), class = c("C", "B", "A"))
}
```

Since classes are just represented by strings, we are not really explicitly defining them; we just have three functions that create objects with the class attribute set. Functions that create objects are known as *constructors* and for the S3 system, this is really the only way of defining classes.

We can create instances of these and check their classes.

```
x <- A()
y <- B()
z <- C()
```

```
class(x)
## [1] "A"
class(y)
## [1] "B" "A"
class(z)
## [1] "C" "B" "A"
```

If we define a generic function that only has a default implementation

```
f <- function(x) UseMethod("f")
f.default <- function(x) print("f.default")
```

calling it on all three objects will just give us the default behavior.

```
f(x)
## [1] "f.default"
f(y)
## [1] "f.default"
f(z)
## [1] "f.default"
```

If, on the other hand, we define a generic function that has a default implementation and an implementation for the class A, calling it on the three objects all results in the A function being called.

```
g <- function(x) UseMethod("g")
g.default <- function(x) print("g.default")
g.A <- function(x) print("g.A")

g(x)
## [1] "g.A"
g(y)
## [1] "g.A"
g(z)
## [1] "g.A"
```

For x we know why this happens. R looks for g.A and finds it. For y, on the other hand, R looks for g.B and doesn't find it. If the class of y was just "B" that would be the end of the search and R would call g.default. But the class of y is both "B" and "A", so when R doesn't find g.B it instead searches for g.A, which does exist, and invokes that. Similarly, when we call g(z), R first searches for g.C, doesn't find it, and so it then searches for g.B, which it also doesn't find. It then finally finds that g.A exists and it calls that. So all three calls are to the same g.A function. Classes B and C inherited the implementation of g from class A.

We can now try adding a generic function with implementations for both class A and B.

```
h <- function(x) UseMethod("h")
h.default <- function(x) print("h.default")
h.A <- function(x) print("h.A")
h.B <- function(x) print("h.B")

h(x)
## [1] "h.A"
h(y)
## [1] "h.B"
h(z)
## [1] "h.B"
```

In this case, calling h(x) invokes h.A—naturally, the only class for x is A and h.A is implemented—while calling h(y) and h(z) both invokes h.B. In both cases, class B is found before class A in the class sequence, and since h.B exists it is invoked, and the search for matching functions stops. The h.A function exists, but we stop searching after the first match, so we never get to it.

For completeness, if we add a function that has implementations for all the three classes, the most specialized function will be called for each.

```
i <- function(x) UseMethod("i")
i.default <- function(x) print("i.default")
i.A <- function(x) print("i.A")
i.B <- function(x) print("i.B")
i.C <- function(x) print("i.C")

i(x)
## [1] "i.A"
i(y)
## [1] "i.B"
i(z)
## [1] "i.C"
```

Getting the Hierarchy Correct in the Constructors

When we created the three classes A, B, and C above we explicitly created the class list representing the class hierarchy in each constructor. This is fine for a small hierarchy like the one we have here, where we define all three classes close together and know that they are supposed to work together, but it does introduce a potential source of errors if the code grows into something more complex. For example, we might at some point want to put another class into the hierarchy between A and B. We might remember to update the class list for B, but will we also remember to update it for C? If C is in a different file, perhaps, and we haven't modified it in years?

Since the class hierarchy is entirely represented in these class lists—there is no explicit hierarchy, but only these implicit lists—we must be extra careful to ensure that our code always matches our design. We should avoid explicitly representing the hierarchy in every constructor.

One way to ensure this is to always invoke the immediate superclass constructor when you define a new class. R doesn't know anything about class hierarchies, and it doesn't know which superclass you have until after you have created the class list. Consequently, there is no automatic way of doing this; you must manually call the constructor of the superclass, obtain an object this way, and then prepend the new class name to its class list.

```
A <- function() {
  structure(list(), class = "A")
}

B <- function() {
  this <- A()
  class(this) <- c("B", class(this))
  this
}

C <- function() {
  this <- B()
  class(this) <- c("C", class(this))
  this
}
```

Using this programming idiom also saves you from another potential error source. The constructor of a class often creates an object with attributes that it expects to be in a consistent state in its functions. If you do not call the constructor, but set up these attributes in the constructor of a subclass, you might violate invariants the superclass depends upon. You will probably be careful not to do that when you write the subclass constructor, but the superclass implementation could be changed in the future. When that happens, the subclass implementation could be broken. It is hell to try to debug an error that shows up because the code has changed in an entirely different location in the code.

You are generally better off if you always call the constructor of the superclass and then modify the resulting object before you return the object of the subclass.

NextMethod

When we specialize generic functions, we do not always need to implement everything from scratch either. Sometimes we can reuse implementations from more abstract classes; we just need to tweak the results of them a little. We can always call the function of a superclass explicitly using the full name of the function, but this will only work if the class actually implements its own version. It will not work if it just inherits it, and you have to be careful every time you modify the class hierarchy that you are not breaking assumptions underlying such direct calls.

A better solution is to use the NextMethod function. This function lets you call inherited functions in a way that resembles UseMethod, and that uses the class sequence. The NextMethod function is magical in a similar way to UseMethod. It knows which generic function you are implementing—although, as with UseMethod you can override this—and it will search through the class sequence of the first object of your function. Again, you can override this, but it is such a common idiom that you probably shouldn't. Unlike UseMethod, though, it doesn't terminate the function you call it from, and you can use the value it returns.

As an example, we can consider the bibliography objects we discussed at the end of the previous chapter. Here we had the superclass "publication" and the two subclasses "article" and "book", and we had a generic method, format, for formatting citations. We can write a format function for the publication class that formats the name and the authors' list, and in the specializations of the method, we can call NextMethod to obtain this string and extend it with information available in the sub-classes.

```
format.publication <- function(publ) {
  paste(name(publ), authors(publ), sep = ", ")
}

format.article <- function(publ) {
  paste(NextMethod(), journal(publ), pages(publ), sep = ", ")
}

format.book <- function(publ) {
  paste(NextMethod(), publisher(publ), ISBN(publ), sep = ", ")
}
```

When we call NextMethod, R will search through the class sequence— starting where it left off last time it did the search—for a class that implements the generic function we are currently evaluating.

We can use the A, B, and C classes to see this in action. If we define functions that all call NextMethod and evaluate it for objects of each of the three classes, we will see that the A object, x, will evaluate the A method and the default function; the B object, y will evaluate the B function and then continue with the A function before the default version; and the C object, z, will evaluate all the functions.

```
f <- function(x) UseMethod("f")
f.default <- function(x) print("f.default")
f.A <- function(x) {
```

```
  print("f.A")
  NextMethod()
}
f.B <- function(x) {
  print("f.B")
  NextMethod()
}
f.C <- function(x) {
  print("f.C")
  NextMethod()
}

f(x)
## [1] "f.A"
## [1] "f.default"
f(y)
## [1] "f.B"
## [1] "f.A"
## [1] "f.default"
f(z)
## [1] "f.C"
## [1] "f.B"
## [1] "f.A"
## [1] "f.default"
```

If we implement another function, where we only implement version for classes A and C but not for B, we will see the C objects evaluating the C version and then the A version; since there is no B version, the search when we call NextMethod will not find it and will thus continue to the A version.

```
j <- function(x) UseMethod("j")
j.default <- function(x) print("j.default")
j.A <- function(x) {
  print("j.A")
  NextMethod()
}
j.C <- function(x) {
  print("j.C")
  NextMethod()
}

j(x)
## [1] "j.A"
## [1] "j.default"
```

```
j(y)
## [1] "j.A"
## [1] "j.default"
j(z)
## [1] "j.C"
## [1] "j.A"
## [1] "j.default"
```

All the method dispatching—the technical term for this looking up of implementations of generic functions—is entirely based on the list of classes in the class attribute. Class hierarchies are not explicitly represented and it is your own responsibility that the class lists reflect the design of class hierarchies you have in mind. If we reverse the order of the class list, R is just as happy to work with this.

```
f(z)
## [1] "f.C"
## [1] "f.B"
## [1] "f.A"
## [1] "f.default"
class(z) <- rev(class(z))
f(z)
## [1] "f.A"
## [1] "f.B"
## [1] "f.C"
## [1] "f.default"
```

If we reverse the order of the classes, the search for the first function in UseMethod just starts at the most abstract class and works its way down the list until it finds an implementation. In this case, it finds it right away, and the calls to NextMethod just continue searching from where UseMethod or previous NextMethod calls ended, exactly like before. The default function is always last here because once you get through the list, the default is invoked. If the A class is an abstract class that C specializes, this will, of course, give you incorrect behavior. You would invoke the abstract and not the specialized function. But the mechanism for resolving generic functions into concrete classes does not know about class hierarchies, so it will just search through the list.

This also means that you can write different classes that consider the hierarchy of other classes to be completely different. If one class list is c("A", "B", "C") and another is c("B", "C", "A"), the first considers C the most abstract and A the most derived class, while the second considers A the most abstract and B the most derived. The mechanism for how R finds the function to call is simple enough, but if you begin writing code where your class hierarchy depends on which object you are looking at, you are on the road to madness.

CHAPTER 4

■ ■ ■

Statistical Models

Truth be told, you won't be using object-oriented programming in most day-to-day R programming. Most analyses you do in R involve the transformation of data, typically implemented as some sort of data flow, which is best captured by functional programming. When you write such pipelines, you will probably be using polymorphic functions, but you rarely need to create your own classes. If you need to implement a new statistical model, however, you usually do want to create a new class.

A lot of data analysis requires that you infer parameters of interest or you build a model to predict properties of your data, but in many of those cases, you don't necessarily need to know exactly how we constructed the model, how it infers parameters, or how it predicts new values. You can use the coefficients function to get inferred parameters from a fitted model, or you can use the predict function to predict new values, and you can use those two functions with almost any statistical model. That is because most models are implemented as classes with implementations for the generic functions coefficients and predict.

As an example of object-oriented programming in action, we can implement our own model in this chapter. We will keep it simple, so we can focus on the programming aspects and not the statistical theory, but still implement something that isn't already built into R. We will implement a version of Bayesian linear regression.

Bayesian Linear Regression

The simplest form of linear regression fits a line to data points. Imagine we have vectors x and y, and we wish to produce coefficients w[1] and w[2] such that y[i] = w[1] + w[2] x[i] + e[i] where the e is a vector of errors that we want to make as small as possible. We typically assume that the errors are identically normally distributed when we consider it a statistical problem, and so we want to have the minimal variance of the errors. When fitting linear models with the lm function, you get the maximum likelihood values for the weights w[1] and w[2], but if you wish to do Bayesian statistics you should instead consider this weight vector w as a random variable, and fit it to the data in x, while y means updating it from its prior distribution to its posterior distribution.

© Thomas Mailund 2017

T. Mailund, *Advanced Object-Oriented Programming in R*, DOI 10.1007/978-1-4842-2919-4_4

A typical distribution for linear regression weights is the normal distribution. If we consider the weights' multivariate normal distributed as their prior distribution, then their posterior distribution given the data will also be normally distributed, which makes the mathematics very convenient.

We will assume that the prior distribution of w is a normal distribution with mean zero and independent components, so a diagonal covariance matrix. This means that, on average, we believe the line we are fitting to be flat and going through the plane origin, but how strongly we believe this depend on values in the covariance matrix. This we will parameterize with a so-called *hyperparameter*, a, that is the precision—one over the variance—of the weight components. The covariance matrix will have 1/a on its diagonal and zeros off-diagonal.

We can represent a distribution over weights as the mean and covariance matrix of a multinomial normal distribution and construct the prior distribution from the precision like this:

```
weight_distribution <- function(mu, S) {
  structure(list(mu = mu, S = S), class = "wdist")
}

prior_distribution <- function(a) {
  mu = c(0, 0)
  S = diag(1/a, nrow = 2, ncol = 2)
  weight_distribution(mu, S)
}
```

We give the weights distribution a class, just to distinguish them from plain lists, but otherwise, there is nothing special to see here.

If we wish to sample from this distribution, we can use the mvrnorm function from the MASS package.

```
sample_weights <- function(n, distribution) {
  MASS::mvrnorm(n = n,
                mu = distribution$mu,
                Sigma = distribution$S)
}
```

We can try to sample some lines from the prior distribution and plot them. We can, of course, plot the sample w vectors as points in the plane, but since they represent lines, we will display them as such (see Figure 4-1).

```
prior <- prior_distribution(1)
(w <- sample_weights(5, prior))
##              [,1]        [,2]
## [1,]   0.6029080 -0.84085548
## [2,]   0.4721664  1.38435934
## [3,]   0.6353713 -1.25549186
## [4,]   0.2857736  0.07014277
## [5,]  -0.1381082  1.71144087
plot(c(-1, 1), c(-1, 1), type = 'n',
     xlab = '', ylab = '')
plot_lines <- function(w) {
  for (i in 1:nrow(w)) {
    abline(a = w[i, 1], b = w[i, 2])
  }
}
plot_lines(w)
```

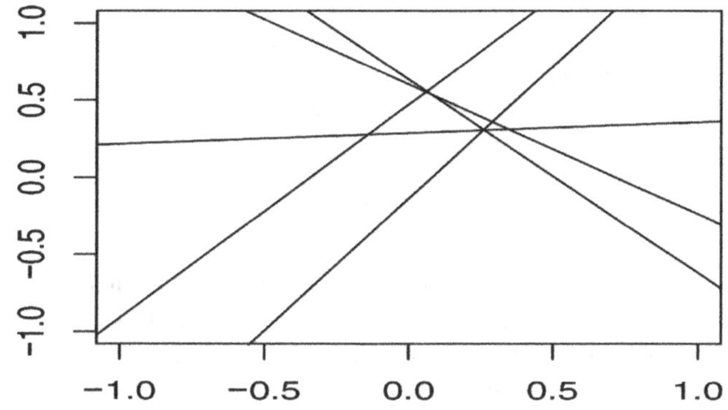

Figure 4-1. *Samples from the prior of lines*

When we observe data in the form of matching X and y values, we must update the w vector to reflect this, which means updating the distribution of the weights. I won't derive the math (this is not a math textbook, after all), but if mu0 is the prior mean and S0 the prior covariance matrix, then the posterior mean and covariance matrix are computed thus:

```
S <- solve(S0 + b * t(X) %*% X)
mu <- S %*% (solve(S0) %*% mu0 + b * t(X) %*% y)
```

45

You cannot execute the code as it is right here—it depends on variables we haven't defined—but it is the body of the function we define shortly. It is a little bit of linear algebra that involves the prior distribution and the observed values. The parameters we haven't seen before in these expressions are b and X. The former is the precision of the error terms—which we assume to know and represent as this hyperparameter—and the latter captures the x values. We cannot use x alone because we want to use two weights to represent lines. When we write w[1] + w[2] * x[i] for the estimate of y[i], we can think of it as the vector product of w and c(0,x[i]), which is exactly what we do. We represent all x[i] as rows c(0,x[i]) in the matrix X. So we estimate y[i] = w[1] * X[i, 1] + w[2] * X[I, 2] in this notation, or y = X %*% w.

As a fitting function, we can write it as this:

```
fit_posterior <- function(x, y, b, prior) {
  mu0 <- prior$mu
  S0 <- prior$S

  X <- matrix(c(rep(1, length(x)), x), ncol = 2)

  S <- solve(S0 + b * t(X) %*% X)
  mu <- S %*% (solve(S0) %*% mu0 + b * t(X) %*% y)

  weight_distribution(mu = mu, S = S)
}
```

We can try to plot some points, fit the model to these, and then plot lines sampled from the posterior. These should fall around the points now, unlike the lines sampled from the prior. The more points we use to fit the model, the tighter that lines sampled from the posterior will fall around the points (see Figure 4-2).

```
x <- rnorm(20)
y <- 0.2 + 1.3 * x + rnorm(20)
plot(x, y)

posterior <- fit_posterior(x, y, 1, prior)
w <- sample_weights(5, posterior)
plot_lines(w)
```

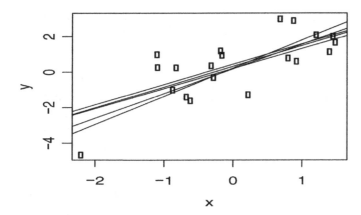

Figure 4-2. *Samples of posterior lines*

The way we update the distribution of weights here from the prior distribution to the posterior doesn't require that the prior distribution is the exact one we created using the variable a. Any distribution can be used as the prior, and in Bayesian statistics, it would not be unusual to update a posterior distribution as more and more data is added. We can start with the prior we just created and fit it to some data to get a posterior. Then if we get more data, we can use the first posterior as a prior for fitting more data and getting a new posterior that captures the knowledge we have gained from seeing all the data. If we have all the data from the get-go, there is no particular benefit to doing this, but in an application where data comes streaming, we can exploit this to quickly update our knowledge each time new data is available.

Since this book is not about Bayesian statistics, and since we only use Bayesian linear regression as an example of writing a new statistical model, we will not explore this further here.

Model Matrices

The X matrix we used when fitting the posterior is an example of a so-called model matrix or design matrix. Using a matrix this way, fitting a response variable, y, to some expression, here 1 + x (in a sense, we used the rows c(1, x[i])), is a general approach to fitting data. Linear models are called *linear* not because we fit data with a line, but because the weights, the w vector we fit, are used linearly, in the mathematical sense, in the fitted model. If we have fitted the weights to the vector w, the line we have fitted is given by X %*% w. That is, our line is the matrix product of the model matrix and the weight vector. The result is a line because X has the form we gave it, but it doesn't have to represent a line for the model to be a linear model. We can transform the input data (here our vector x) in any way we want to before we fit the model. You might have used log-transformations before, or fitted data to a polynomial, and those would also be examples of linear models.

47

You can fit various kinds of transformed data using the function we wrote above if you just construct the X matrix in different ways. As long as each data point you have becomes a row in X, it doesn't matter what you do. The same mathematics work. To fit a quadratic equation to the data instead of a line, you just have to make the ith row of X be c(1, x[i], x[i]**2), for example.

In R, you have a very powerful mechanism for constructing model matrices from formulae. Whenever you have used a formula such as y ~ x + y to fit y to two variables, x and z, you have used this feature. The formula is translated into a model matrix, and once you have the matrix, the code that does the model fitting doesn't need to know anything else about your data.

The function you use to translate a formula into a model matrix is called model.matrix. Even though it has a period in its name, it isn't a generic function. It just has an unfortunate name for historical reasons.

The function will take a formula as input and produce a model matrix. It will find the values for the variables in the formula by looking in its scope, and we can use it like this:

```
x <- rnorm(5)
y <- 1.2 + 2 * x + rnorm(5)
model.matrix(y ~ x)
##   (Intercept)          x
## 1           1 -1.1365828
## 2           1  0.8548304
## 3           1 -0.5783704
## 4           1  0.4963615
## 5           1 -0.7600579
## attr(,"assign")
## [1] 0 1
```

Relying on global variables like this is risky coding, though, so we often put our data in a data frame instead.

```
d <- data.frame(x, y)
```

If we do this, we can provide a data frame to the model.matrix function and it will get the variables from there.

```
model.matrix(y ~ x, data = d)
##   (Intercept)          x
## 1           1 -1.1365828
## 2           1  0.8548304
## 3           1 -0.5783704
## 4           1  0.4963615
## 5           1 -0.7600579
## attr(,"assign")
## [1] 0 1
```

If the formula uses variables that are not found in the data frame, the model.matrix will still find the missing variables in the calling scope, but this is really risky coding, so you should avoid this.

The formula determines how the model matrix is constructed. The formula y ~ x gives us the model matrix we used above, where we have 1 to capture the intercept in the first column, and we have the x values in the second column to capture the incline of the line. We can remove the intercept using the formula y ~ x - 1:

```
model.matrix(y ~ x - 1, data = d)
##            x
## 1 -1.1365828
## 2  0.8548304
## 3 -0.5783704
## 4  0.4963615
## 5 -0.7600579
## attr(,"assign")
## [1] 1
```

We can also add terms; for example, we can fit a quadratic formula to the data by constructing this model matrix:

```
model.matrix(y ~ x + I(x**2), data = d)
##   (Intercept)         x      I(x^2)
## 1           1 -1.1365828 1.2918205
## 2           1  0.8548304 0.7307351
## 3           1 -0.5783704 0.3345123
## 4           1  0.4963615 0.2463748
## 5           1 -0.7600579 0.5776881
## attr(,"assign")
## [1] 0 1 2
```

Here, we need to wrap the squared x in the function I to get R to use the actually squared values of x. Inside formulae, products are interpreted as interaction, but by wrapping x**2 in I, we make it the square of the x values explicitly.

The model matrix doesn't include the response variable, y, so we cannot get that from it. Instead, we can use a related function, model.frame, that also gives us a column for the response.

```
model.frame(y ~ x + I(x**2), data = d)
##              y              x        I(x^2)
## 1 -1.4145519 -1.1365828 1.291820....
## 2  0.8073317  0.8548304 0.730735....
## 3 -0.2584431 -0.5783704 0.334512....
## 4  0.9203396  0.4963615 0.246374....
## 5 -0.5997820 -0.7600579 0.577688....
```

You can then extract the response variable using the function model.response, like this:

```
model.response(model.frame(y ~ x + I(x**2), data = d))
##            1            2            3            4
## -1.4145519    0.8073317   -0.2584431    0.9203396
##            5
## -0.5997820
```

With that machinery in place, we can generalize our distributions and model fitting to work with general formulae. We can write the prior function like this:

```
prior_distribution <- function(formula, a, data) {
  n <- ncol(model.matrix(formula, data = data))
  mu <- rep(0, n)
  S <- diag(1/a, nrow = n, ncol = n)
  weight_distribution(mu, S)
}
```

Ideally, a prior shouldn't depend on any data, but the form of a model matrix does depend on the type of the data we use. Numerical data will be represented as a single column in the model matrix, but factors are handled as a binary vector for each level in a formula, so we do need to know what kind of data we are going to need. We could have added n as a parameter here and made the function independent of any data, but I chose to include a data frame as another solution. We don't use the actual data, though; we just use it to get the number of columns in the model frame.

The function for fitting the data changes less, though. It just uses the model.matrix function to construct the model matrix instead of the explicit construction we did earlier.

```
fit_posterior <- function(formula, b, prior, data) {
  mu0 <- prior$mu
  S0 <- prior$S

  X <- model.matrix(formula, data = data)
```

```
  S <- solve(S0 + b * t(X) %*% X)
  mu <- S %*% (solve(S0) %*% mu0 + b * t(X) %*% y)

  weight_distribution(mu = mu, S = S)
}
```

With that in place, we can fit a line as before, but now using a formula.

```
d <- {
  x <- rnorm(5)
  y <- 1.2 + 2 * x + rnorm(5)
  data.frame(x = x, y = y)
}
```

```
prior <- prior_distribution(y ~ x, 1, d)
posterior <- fit_posterior(y ~ x, 1, prior, d)
posterior
## $mu
##                   [,1]
## (Intercept) 1.1979166
## x           0.2875647
##
## $S
##                (Intercept)          x
## (Intercept)   0.16935519 -0.04441203
## x            -0.04441203  0.73364645
##
## attr(,"class")
## [1] "wdist"
```

If we, instead, want to fit a quadratic function, we do not need to change any of the functions, we can just provide a different formula.

```
prior <- prior_distribution(y ~ x + I(x**2), 1, d)
posterior <- fit_posterior(y ~ x + I(x**2), 1, prior, d)
posterior
## $mu
##                   [,1]
## (Intercept) 1.1906733
## x           0.2815418
## I(x^2)      0.1185518
##
## $S
```

```
##              (Intercept)            x        I(x^2)
## (Intercept)  0.17301554 -0.04136841 -0.05990922
## x           -0.04136841  0.73617725 -0.04981521
## I(x^2)      -0.05990922 -0.04981521  0.98053972
##
## attr(,"class")
## [1] "wdist"
```

Constructing Fitted Model Objects

Now, we want to wrap fitted models in a class so we can write a constructor for them. For fitted models, it is traditional to include the formula, the data, and the function call together with the fitted model, so we will put those as attributes in the objects. The constructor could look like this:

```
blm <- function(formula, b, data, prior = NULL, a = NULL) {

  if (is.null(prior)) {
    if (is.null(a)) stop("Without a prior you must provide a.")
    prior <- prior_distribution(formula, a, data)
  } else {
    if (inherits(prior, "blm")) {
      prior <- prior$prior
    }
  }
  if (!inherits(prior, "wdist")) {
    stop("The provided prior does not have the expected type.")
  }

  posterior <- fit_posterior(formula, b, prior, data)

  structure(
    list(formula = formula,
         data = model.frame(formula, data),
         dist = posterior,
         call = match.call()),
    class = "blm"
  )
}
```

The tests at the beginning of the constructor allow us to specify the prior as either a normal distribution or a previously fitted blm object. If we get a prior distribution, we probably should also check that this prior is compatible with the actual formula, but I will let you write such a check yourself to try that out.

If we print objects fitted using the blm function, they will just be printed as a list, but we can provide our own print function by specializing the print generic function. Here, it is tradition to provide the function call used to specify the model, so that is all I will do for now.

```
print.blm <- function(x, ...) {
  print(x$call)
}
```

We have to match the parameters of the generic function, which is why the arguments are x and.... All we do, however, is print the call attribute of the object.

Now, we can fit data and get a blm object like this:

```
(model <- blm(y ~ x + I(x**2), a = 1, b = 1, data = d))
## blm(formula = y ~ x + I(x^2), b = 1, data = d, a = 1)
```

Coefficients and Confidence Intervals

Once we have a fitted model, we might want to get the fitted values. This is traditionally done using the generic function coef, which should simply return those. For the Bayesian linear regression model, the fitted values are whole distributions, but we can take the mean values as point estimates and return those, and so implement the coef function like this:

```
coef.blm <- function(object, ...) {
  t(object$dist$mu)
}
coef(model)
##      (Intercept)        x     I(x^2)
## [1,]   1.190673 0.2815418 0.1185518
```

Here, I transform the mean matrix to get the coefficients in a form similar to what you would get with a traditional linear model, as fitted with the function lm, but other than that, the function just returns the means.

For classical frequentist models, we often also want the confidence intervals for the parameters, and the traditional way to get those is using the confint function. This function has the signature:

```
confint(object, parm, level = 0.95, ...)
```

The object parameter is the fitted model, and the parm contains the parameters we want the confidence intervals for. If it is missing we should return all the model parameters, and the level parameter specifies at which confidence levels we want the intervals.

Again, for a Bayesian model, we have whole distributions and not just confidence intervals, but we can get something similar for our model by getting the quantiles from the marginal distributions for each parameter. The parameters are normally distributed, and we can get the means and standard deviations from the means and covariance matrix, respectively. After that, we can get the quantiles using the qnorm function and construct intervals like this:

```
confint.blm <- function(object, parm, level = 0.95, ...) {
  if (missing(parm)) {
    parm <- rownames(object$dist$mu)
  }

  means <- object$dist$mu[parm,]
  sds <- sqrt(diag(object$dist$S)[parm])

  lower_q <- qnorm(p = (1-level)/2,
                   mean = means,
                   sd = sds)
  upper_q <- qnorm(p = 1 - (1-level)/2,
                   mean = means,
                   sd = sds)

  quantiles <- cbind(lower_q, upper_q)
  quantile_names <- paste(
    100 * c((1-level)/2, 1 - (1 -level)/2),
    "%",
    sep = ""
    )
  colnames(quantiles) <- quantile_names

  quantiles
}

confint(model)
##                  2.5%      97.5%
## (Intercept)  0.3754236 2.005923
## x           -1.4001225 1.963206
## I(x^2)      -1.8222478 2.059351
```

Predicting Response Variables

When it comes to predicting response variables for new data, there is a little more work to be done for the model matrix. If we don't have the response variable when we create a model matrix, we will get an error, even though the model matrix doesn't actually contain a column with it. We can see this if we remove the x and y variables we used when creating the data frame earlier.

```
rm(x) ; rm(y)
```

We are fine if we build a model matrix from the data frame that has both x and y.

```
model.matrix(y ~ x + I(x**2), data = d)
##   (Intercept)          x        I(x^2)
## 1           1 -0.20409732 0.041655716
## 2           1 -0.22561419 0.050901761
## 3           1  0.34702845 0.120428747
## 4           1  0.03236784 0.001047677
## 5           1  0.41353129 0.171008128
## attr(,"assign")
## [1] 0 1 2
```

If we create a data frame that only has x values, though, we get an error.

```
dd <- data.frame(x = rnorm(5))
model.matrix(y ~ x + I(x**2), data = dd)
## Error in eval(expr, envir, enclos): object 'y' not found
```

This, of course, is a problem since we are interested in predicting response variables exactly when we do not have them. This is ironic, considering that the model matrix never actually contains the response variable; nevertheless, this is how R works.

To remove the response variable from a formula, before we construct a model matrix, we need to use the function delete.response. Unfortunately, this function does not work directly on formulae, but on so-called "terms" objects. We can translate a formula into its terms using the terms function. So to remove the terms from a formula, we need the following construction:

```
delete.response(terms(y ~ x))
```

The result is a terms object, but since the terms class is a specialization of the formula class, we can use it directly with the model.matrix function to construct the model matrix.

```
model.matrix(delete.response(terms(y ~ x)), data = dd)
##    (Intercept)              x
## 1            1   0.49841617
## 2            1  -1.74230249
## 3            1   0.97552910
## 4            1  -0.02408287
## 5            1   0.67568448
## attr(,"assign")
## [1] 0 1
```

Now, for actually predicting data, we traditionally use the predict generic function. This function has the following signature:

```
function(object, ...)
```

Which, quite frankly, isn't that useful. We can add to it, though, as the lm class does, so we can add a parameter newdata to it. From newdata we will construct a model matrix and make predictions for the data in this. If we just want point estimates for the new data, we can simply take the inner product of the mean weights and each row in the model matrix so we can implement the predict function like this:

```
predict.blm <- function(object, newdata, ...) {
  updated_terms <- delete.response(terms(object$formula))
  X <- model.matrix(updated_terms, data = newdata)

  predictions <- vector("numeric", length = nrow(X))
  for (i in seq_along(predictions)) {
    predictions[i] <- t(object$dist$mu) %*% X[i,]
  }
  predictions
}
```

```
predict(model, d)
## [1] 1.138150 1.133188 1.302653 1.199910 1.327373
```

To check the model, we can plot the predicted values against the true response values. With the data we have used up till now, though, with only five data points, we don't expect the predictions to be particularly good, and we don't expect such a plot to show much, but we can try with a few more data points (see Figure 4-3).

```
d <- {
  x <- rnorm(50)
  y <- 0.2 + 1.4 * x + rnorm(50)
  data.frame(x = x, y = y)
}
model <- blm(y ~ x, d, a = 1, b = 1)
plot(d$y, predict(model, d),
     xlab = "True responses",
     ylab = "Predicted responses")
```

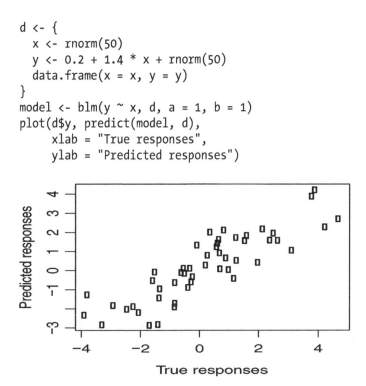

Figure 4-3. *True versus predicted values*

Predicting values for the original data is so common that there is another generic function for doing this, called fitted. We can implement it like this:

```
fitted.blm <- function(object, ...) {
  predict(object, newdata = object$data, ...)
}
```

With this function, we could then create the plot as follows:

```
plot(d$y, fitted(model),
     xlab = "True responses",
     ylab = "Predicted responses")
```

We can, of course, do more than simply predict point estimates. We have a distribution of weights, which means that the slope and intercept of a line aren't fixed. They are, conceptually at least, drawn from a distribution. Slight changes in the slope won't have much of an impact on how certain we are in the predictions near the main mass of the data, but data points further towards the edges of the data will have a larger uncertainty because of this. We can take the distribution into account when we make predictions to get error bars for the predictions. If X is the model matrix and S the covariance matrix for the posterior, then the variance for the ith prediction is given by:

```
1/b + t(X[i,]) %*% S %*% X[i,]
```

We don't have the precision parameter, b, stored in the fitted model object, so if we want to get error bars, we need to keep that around. So update your blm function to return this structure:

```
structure(
  list(formula = formula,
       data = model.frame(formula, data),
       dist = posterior,
       precision = b,
       call = match.call()),
  class = "blm"
)
```

Now, we can extend our predict function with an option that, if TRUE, returns intervals as well:

```
predict.blm <- function(object, newdata,
                        intervals = FALSE,
                        level = 0.95,
                        ...) {

  updated_terms <- delete.response(terms(object$formula))
  X <- model.matrix(updated_terms, data = newdata)

  predictions <- vector("numeric", length = nrow(X))
  for (i in seq_along(predictions)) {
    predictions[i] <- t(object$dist$mu) %*% X[i,]
  }

  if (!intervals) return(predictions)

  S <- model$dist$S
  b <- model$precision
  sds <- vector("numeric", length = nrow(X))
  for (i in seq_along(predictions)) {
    sds[i] <- sqrt(1/b + t(X[i,]) %*% S %*% X[i,])
  }
```

```
    lower_q <- qnorm(p = (1-level)/2,
                     mean = predictions,
                     sd = sds)
    upper_q <- qnorm(p = 1 - (1-level)/2,
                     mean = predictions,
                     sd = sds)

    intervals <- cbind(lower_q, predictions, upper_q)
    colnames(intervals) <- c("lower", "mean", "upper")
    as.data.frame(intervals)
}

model <- blm(y ~ x, d, a = 1, b = 1)
```

With this `predict` function, we can plot error bars around our predictions as well, see Figure 4-4. In the figure, I also plot the $x=y$ line. If we make accurate predictions, the point should lie on this line. Since there is some stochasticity in the data, we don't expect to be exactly on it, but we do expect to overlap the line 95% of the time.

```
require(ggplot2)
predictions <- fitted(model, intervals = TRUE)
ggplot(cbind(data.frame(y = d$y), predictions),
       aes(x = y, y = mean)) +
  geom_point() +
  geom_errorbar(aes(ymin = lower, ymax = upper)) +
  geom_abline(slope = 1) +
  xlab("True responses") +
  ylab("Predictions") +
  theme_minimal()
```

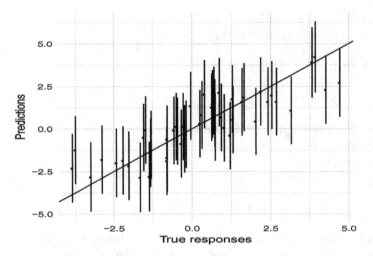

Figure 4-4. *Predictions versus true values with confidence intervals*

There is, of course, much more we can do with a statistical model, and more generic functions we could add to it to provide a uniform interface to a class of models, but I hope this example has at least given you an impression of how representing a model as a class can be useful.

CHAPTER 5

Operator Overloading

Overloading operators, that is, giving operators such as + or - new or additional functionality, is not inherently object-oriented, but since it can be thought of as adding polymorphism to functions it fits in naturally here after we have gone through polymorphism through generic functions.

Opinions vary on whether overloading operators is good or bad practice. Some languages allow it; others do not. Some languages allow you to make your own infix operators but not change existing ones, and some languages are just inconsistent in allowing some operator overloading for built-in objects but not for user-defined. One argument against overloading is the expected behavior of operators. It is so ingrained in us to expect + to mean addition that we cannot handle if it also means string concatenation. This argument, of course, ignores that we have no problem with using + for both integer and floating point addition. Another argument, and in my opinion a more valid one, is that it is harder to remember what an operator does than what a function does since the function name at least gives us some hint as to the function's purpose. The truth is that infix operators, when carefully used, can provide us with a more convenient syntax than simple function calls. You probably find `2 * x + 5` easier to read than `plus(times(2, x), 5)`, and most programming languages, with Lisp dialects being a noticeable exception, prefer the former to the latter. The same goes for user-defined or user-overloaded operators. Using `magrittr`'s pipe operator, `%>%`, makes analysis workflows much easier to read, and `ggplot2`'s overloading of + makes plotting code much easier to read.

Since, in R, you can both overload existing operators and create your own, you have to choose which is most appropriate for any given situation. My rule of thumb is to prefer new infix operators unless it feels natural to use an existing one. That is, of course, a terribly subjective evaluation, but some operations just seem like "addition"—I would say combining operations in `ggplot2` can be justified as such—while others don't, like pipeline operations in `magritte`. It is a judgment call, and you can always experiment with your code to see what you feel is most natural.

In this chapter, we will see how we can overload operators. I will not cover how you create infix operators. If you are interested, I describe those in my *Functional Programming in R* book.

© Thomas Mailund 2017
T. Mailund, *Advanced Object-Oriented Programming in R*, DOI 10.1007/978-1-4842-2919-4_5

Functions and Operators

Every operation in R involves a function call. Control structures, subscripting, even parenthesis involve functions, and, naturally, operators involve function calls as well. This means that operators can be overridden. You can replace one implementation of + with another just by defining a new version of +. But you really shouldn't. Replacing the built-in operators with user-defined ones will affect all your code, slow it down, and is very likely to introduce hard-to-find bugs. You don't have to do this to define operators for your own classes, though. The operators are generic, and you can define specialized versions of them, implementing how any operator should handle new classes you define.

To illustrate how this is done, we define a class for arithmetic modulus some number n. Here I assume that value is some numeric type—in production code, we would write tests for this, but in the example, we just implicitly assume this—and we use an attribute to store the number n. We then compute the value modulus n, set the class, and return the result.

```
modulus <- function(value, n) {
  result <- value %% n
  attr(result, "modulus") <- n
  class(result) <- c("modulus", class(value))
  result
}
```

To pretty-print values of the class, we define the print method. We want to print values, x, using the underlying type but with a line above giving us n. If we just call print(x) we would recurse back to print.modulus since that is the type of x, but we can use the function unclass to get rid of the type of x. It doesn't completely get rid of the type, though. If x is a numeric type, unclass just reduces x to that. When we print primitive values with attributes the attributes are printed as well, so we get rid of all attributes before we print the reduced x. Here we could directly call print(x) but NextMethod() works as well, so that is what I have used here.

```
print.modulus <- function(x, ...) {
  cat("Modulus", attr(x, "modulus"), "values:\n")
  # remove attributes to get plain numeric printing
  x <- unclass(x)
  attributes(x) <- NULL
  NextMethod()
}

(x <- modulus(1:6, 3))
## Modulus 3 values:
## [1] 1 2 0 1 2 0
```

Defining Single Operators

We can define what addition means for this type by defining the function

`` `+.modulus` ``:

```
`+.modulus` <- function(x, y) {
  n <- attr(x, "modulus")
  x <- unclass(x)
  y <- unclass(y)
  modulus(x + y, n)
}
```

We first get n from x. Then we remove the class information from the two operands so we can call the primitive + to calculate their sum. If we didn't do this, we would again recurse when we tried to add x to y, and we would then simply calculate the result and use the modulus constructor to return it with the right type.

Dispatching of such operator-functions works a little different from the generic functions we have seen earlier. There, the dispatch is based on the type of the first argument, at least unless you explicitly state otherwise, but for operators, the dispatch is based on the type of both operands. If both have a class that implements an operator, they must have the same class. If only one of them have a class, and here primitive classes such as "numeric" or "integer" do not count, then the dispatch is based on that. So if x is of type "modulus" then both x + 1:6 and 1:6 + x will call `` `+.modulus` ``.

```
x + 1:6
## Modulus 3 values:
## [1] 2 1 0 2 1 0
1:6 + x
## Modulus values:
## numeric(0)
```

The first expression does what we expect, but the second does not. This is not because the dispatch is not working but because we got n from the first operand only. Since we are only guaranteed that *one* of the two operands have type "modulus" we need to check both. We can do this simply by checking if the attribute "modulus" is NULL or not:

```
`+.modulus` <- function(x, y) {
  n <- ifelse(!is.null(attr(x, "modulus")),
              attr(x, "modulus"), attr(y, "modulus"))
  x <- unclass(x)
  y <- unclass(y)
  modulus(x + y, n)
}
```

```
x + 1:6
## Modulus 3 values:
## [1] 2 1 0 2 1 0
1:6 + x
## Modulus 3 values:
## [1] 2 1 0 2 1 0
```

Depending on which semantic we want addition of these types to have, we might not want to allow addition of types with different n. If we add such types now, the n is taken from the first operand.

```
y <- modulus(1:6, 2)
x + y
## Modulus 3 values:
## [1] 2 2 1 1 0 0
```

With a little more check we ensure that the two operands are compatible:

```
`+.modulus` <- function(x, y) {
  nx <- attr(x, "modulus")
  ny <- attr(y, "modulus")
  if (!is.null(nx) && !is.null(ny) && nx != ny)
    stop("Incompatible types")
  n <- ifelse(!is.null(nx), nx, ny)

  x <- unclass(x)
  y <- unclass(y)
  modulus(x + y, n)
}

x + y
## Error in `+.modulus`(x, y): Incompatible types
y <- modulus(rev(1:6), 3)
x + y
## Modulus 3 values:
## [1] 1 1 1 1 1 1
```

Group Operators

Using generic functions, we can define all relevant operators for a user-defined type, but it is also possible to handle all operators in a single function, Ops. This function is called a "group generic method" because it handles a group of generic functions; other group methods are Math, Complex, and Summary, which we will not cover here.

If we define the function Ops.modulus, it will be called for all operators of modulus objects where the operator function is not defined. That is, if we have defined `+.modulus` as above, that function will be preferred over Ops.modulus, but otherwise, if one or both of the operands are of type modulus, then Ops.modulus will be called.

We can define it like this:

```
Ops.modulus <- function(e1, e2) {
  nx <- attr(e1, "modulus")
  ny <- attr(e2, "modulus")
  if (!is.null(nx) && !is.null(ny) && nx != ny)
    stop("Incompatible types")
  n <- ifelse(!is.null(nx), nx, ny)

  result <- unclass(NextMethod()) %% n
  modulus(result, n)
}
```

The testing of input values at the beginning of the function is the same as for `+.modulus`. After the testing, we use NextMethod to call the operation using the underlying type. This strips the modulus class from the operands and evaluates whatever operation we are currently handling. We unclass the result, necessary because the result will inherit the attributes of the operands of Ops, so if we don't the result will have class modulus, and we then compute modulus n. If we didn't unclass, this would be a recursive call, but since we do remove the class we just do modulus in the underlying type. We finally create a modulus object of the result.

With this function defined we get all binary operators in one go.

```
y <- modulus(rev(1:6), 3)
x - y
## Modulus 3 values:
## [1] 1 0 2 1 0 2
x * y
## Modulus 3 values:
## [1] 0 1 0 0 1 0
```

This includes comparison operators as well:

```
x == y
## Modulus 3 values:
## [1] 0 1 0 0 1 0
x == x
## Modulus 3 values:
## [1] 1 1 1 1 1 1
```

```
x != y
## Modulus 3 values:
## [1] 1 0 1 1 0 1
x != x
## Modulus 3 values:
## [1] 0 0 0 0 0 0
```

It even includes unary operators. If we use a unary minus, however, the argument e2 will be missing in the function call, which we do not handle correctly right now.

```
- x
## Error in Ops.modulus(x): argument "e2" is missing, with no default
```

We can easily fix this, however, by checking if e2 is missing. Otherwise, the function will work as it is.

```
Ops.modulus <- function(e1, e2) {
  nx <- attr(e1, "modulus")
  ny <- if (!missing(e2)) attr(e2, "modulus") else NULL
  if (!is.null(nx) && !is.null(ny) && nx != ny)
    stop("Incompatible types")
  n <- ifelse(!is.null(nx), nx, ny)

  result <- unclass(NextMethod()) %% n
  modulus(result, n)
}
```

```
- x
## Modulus 3 values:
## [1] 2 1 0 2 1 0
```

Units Example

For a slightly more involved example, we define a class for associating physical units with values. This will allow us to check that units we manipulate are compatible—so we do not subtract meters from seconds and such—and will do unit analysis as part of arithmetic operations. The example is a simplified version of the package units. The units package also handles unit conversion and unit simplification. Here we just implement a simple arithmetic of symbolic units and a simple equality check of them.

The idea is to have a representation of physical units and then associate these to numeric values. Physical units, here, refers to units like square kilometres, metres per second, etc. In general, these will be symbolic expressions, but we will only consider the slightly simpler situation where the units are a fraction of physical constants. In that case, we can represent these as a list of terms in the nominator and a list of terms in the denominator. If we always keep these lists sorted, we have a canonical representation of them, and we can check equality of two units by checking equality of the nominator and denominator lists. We can implement the constructor like this:

```
symbolic_unit <- function(nominator, denominator = "") {
  non_empty <- function(x) x != ""
  nominator <- sort(Filter(non_empty, nominator))
  denominator <- sort(Filter(non_empty, denominator))
  structure(list(nominator = nominator,
                 denominator = denominator),
            class = "symbolic_unit")
}
```

We can translate these units into a string representation of the fraction for pretty-printing.

```
as.character.symbolic_unit <- function(x, ...) {
  format_terms <- function(terms, op) {
    if (length(terms) == 0) return("1")
    paste0(terms, collapse = op)
  }
  nominator <- format_terms(x$nominator, "*")
  denominator <- format_terms(x$denominator, "/")
  paste(nominator, "/", denominator)
}

print.symbolic_unit <- function(x, ...) {
  cat(as.character(x, ...), "\n")
}

(x <- symbolic_unit("m"))
## m / 1
(y <- symbolic_unit("m", "s"))
## m / s
```

Comparing two symbolic units involves checking that the nominator and denominator are equal.

```
`==.symbolic_unit` <- function(x, y) {
  if (!(inherits(x, "symbolic_unit") &&
      inherits(y, "symbolic_unit")))
    stop("Incompatible types")
  return(identical(x$nominator, y$nominator) &&
         identical(x$denominator, y$denominator))
}

`!=.symbolic_unit` <- function(x, y) !(x == y)

x == y
## [1] FALSE
x != y
## [1] TRUE
```

Adding and subtracting physical quantities is only possible if they have the same units, but it is always possible to multiply and divide units. The resulting unit is then obtained by doing the same operation on the (symbolic) units as you do on the quantities. To be able to handle this, we define multiplication and division on symbolic units.

```
`*.symbolic_unit` <- function(x, y) {
  symbolic_unit(c(x$nominator, y$nominator),
                c(x$denominator, y$denominator))
}

`/.symbolic_unit` <- function(x, y) {
  symbolic_unit(c(x$nominator, y$denominator),
                c(x$denominator, y$nominator))
}

x * y
## m*m / s
x / y
## m*s / m
```

We now have everything in place to represent units. We just need to define the class for associating units with quantities. This class is very similar to the modulus class we wrote earlier. We take a (numeric) value, associate a symbolic unit in an attribute, and set the class.

```
units <- function(value, nominator, denominator = "") {
  attr(value, "units") <- symbolic_unit(nominator, denominator)
  class(value) <- c("units", class(value))
  value
}
```

Pretty-printing follows the pattern we saw with modulus. We need to strip the class and attributes to use the underlying print method, called through NextMethod, but that is all there is to it.

```
print.units <- function(x, ...) {
  cat("Units: ", as.character(attr(x, "units")), "\n")
  # remove attributes to get plain numeric printing
  x <- unclass(x)
  attributes(x) <- NULL
  NextMethod()
}

(x <- units(1:6, "m"))
## Units:  m / 1
## [1] 1 2 3 4 5 6
```

Handling operators for units is only slightly more involved than it was for modulus. We need to distinguish between operators where we require that the units match and those where we need to construct new units. The former of those are addition, subtraction, and comparisons, assuming we only want to consider numbers equal if they agree in both quantity and associated units; the latter are multiplication and division where the resulting units must be computed from the operands. It is not obvious how to handle logical operators on physical quantities, if that is something that even makes sense, so for operators that do not fall into these two categories, we should just default to what the underlying type does.

We implement the operators using the Ops group function. Inside this function, we can get hold of the actual operator being evaluated using the variable .Generic. This is not a parameter of the function, but it will be set to the operator being evaluated when the function is called, and we can check the operator and handle it appropriately by switching on it.

```r
Ops.units <- function(e1, e2) {
  su1 <- attr(e1, "units")
  su2 <- if (!missing(e2)) attr(e2, "units") else NULL

  if (.Generic %in% c("+", "-", "==", "!=",
                      "<", "<=", ">=", ">")) {
    if (!is.null(su1) && !is.null(su2) && su1 != su2)
      stop("Incompatible units")
    su <- ifelse(!is.null(su1), su1, su2)
    return(NextMethod())
  }

  if (.Generic == "*" || .Generic == "/") {
    if (is.null(su1))
      su1 <- symbolic_unit("")
    if (is.null(su2))
      su2 <- symbolic_unit("")
    su <- switch(.Generic, "*" = su1 * su2, "/" = su1 / su2)
    result <- NextMethod()
    attr(result, "units") <- su
    return(result)
  }

  # For the remaining operators we don't really have a good
  # way of treating the units so we strip that info and go
  # back to numeric values
  e1 <- unclass(e1)
  e2 <- unclass(e2)
  attributes(e1) <- attributes(e2) <- NULL
  NextMethod()
}
```

With this definition of the units operators we can combine units with scalars:

```r
2 * x
## Units:  m / 1
## [1]  2  4  6  8 10 12
x + 2
## Units:  m / 1
## [1] 3 4 5 6 7 8
x - 2
## Units:  m / 1
## [1] -1  0  1  2  3  4
```

If we attempt to add two quantities with incompatible types we will be warned that this is incorrect.

```
(y <- units(1:6, "m", "s"))
## Units:  m / s
## [1] 1 2 3 4 5 6
x + y
## Error in Ops.units(x, y): Incompatible units
```

When the units are compatible, though, we can add and subtract.

```
(z <- units(1:6, "m"))
## Units:  m / 1
## [1] 1 2 3 4 5 6
x + z
## Units:  m / 1
## [1]  2  4  6  8 10 12
x - z
## Units:  m / 1
## [1] 0 0 0 0 0 0
```

Multiplication and division are always permissible and the resulting units are derived from the operands.

```
2 * x
## Units:  m / 1
## [1]  2  4  6  8 10 12
x * y
## Units:  m*m / s
## [1]  1  4  9 16 25 36
x / y
## Units:  m*s / m
## [1] 1 1 1 1 1 1
```

## CHAPTER 6

S4 Classes

The S3 class system is the wild west of object-oriented programming. Class hierarchies only exist implicitly through the class attributes, and generic methods can be implemented or not by any class whatsoever, with no check that interfaces and class hierarchy designs are implemented correctly. Everything depends on conventions and it is entirely up to the programmer to ensure anything resembling consistency.

The S3 system is popular because it is very easy to use and to write new classes. In most cases, we would prefer simplicity over elaborate design, and in those cases, S3 is perfect for our needs. When we implement a new statistical model, we rarely need a complex class hierarchy. Most abstract data structures have only a few associated operations, and we have no problem remembering to implement them all when we write concrete versions of them. Once software reaches a certain level of complexity, however, more structure is needed. Rather than having the design exist only implicitly in programming conventions we want it explicitly stated in the code.

The S4 system provides a more structured object-oriented system. Here classes and class hierarchies are explicitly created; they are not merely strings in a class attribute. To obtain the added structure that S4 allows, a little more code is needed when creating classes and methods, but overall the system works very similar to S3. If you are comfortable with S3, then, learning S4 should not pose a problem.

Defining S4 Classes

In S4, classes are explicitly created. To create a new class, we use the function setClass from the methods package. This function takes arguments that specify which attributes the objects of the class should hold, what default values the attributes should have, how the class fits into a class hierarchy, and many other properties of the created class. All these properties have default values so we can create a new class just by specifying its name.

As an example, we consider the stack data structure we implemented in S3 earlier. To make an abstract stack class we can write:

```
library(methods)
Stack <- setClass("Stack")
```

This creates a new class called "Stack". We have not specified any attributes of the class, so S4 will assume that it is an abstract class that is not supposed to be instantiated and we will get an error if we try.

We can create the vector-based stack class like this:

```
VectorStack <- setClass("VectorStack",
                slots = c(
                    elements = "vector"
                ),
                contains = "Stack")
```

Here we use two arguments to setClass: slots and contains. The slots argument is a list of attributes that objects of the class should have. Here specify that it should contain a vector called elements. The contains argument specifies which superclasses the new class should have. We make VectorStack a subclass of Stack.

Since VectorStack has slots, it is not considered an abstract class (we can explicitly make it so by adding "VIRTUAL" to the contains argument, but we do want to be able to instantiate it). We can create objects of the class by calling VectorStack.

```
(vs <- VectorStack())
## An object of class "VectorStack"
## Slot "elements":
## logical(0)
```

We can specify the elements as a named argument to VectorStack.

```
(vs <- VectorStack(elements = 1:4))
## An object of class "VectorStack"
## Slot "elements":
## [1] 1 2 3 4
```

Positional arguments will not work here; it has to be a named argument.

Once we have an object of class VectorStack we can access the elements with this notation:

```
vs@elements
## [1] 1 2 3 4
```

In general, @ is used to access slots of S4 objects.

Capturing the result of setClass and using it as a function to construct objects is my preferred way of creating S4 constructors, but strictly speaking, it isn't necessary. Once a class is built with setClass you can create objects just using the name of the class, using the function new.

```
new("VectorStack", elements = 1:4)
## An object of class "VectorStack"
## Slot "elements":
## [1] 1 2 3 4
```

Generic Functions

In the S3 system, we create generic functions just as normal functions that call UseMethod. In S4 we have to define generic functions explicitly using the setGeneric function. The first argument of setGeneric is the name of the generic function. If you have an existing function that you want to make generic, calling setGeneric with its name will create a generic function with the existing function as the default implementation. Typically, though, we use setGeneric to create a brand new generic function, and in that case, we need to provide a definition of the function as well. This we do through the parameter def. This function plays the role that the function definition in S3 has: it should simply call the function standardGeneric, which is S4's analogue to UseMethod.

We can define the interface of the stack abstract data structure like this:

```
setGeneric("top",
           def = function(stack) standardGeneric("top"))
setGeneric("pop",
           def = function(stack) standardGeneric("pop"))
setGeneric("push",
           def = function(stack, element) standardGeneric("push"))
setGeneric("is_empty",
           def = function(stack) standardGeneric("is_empty"))
```

To provide implementations of generic functions we use the function setMethod. We need to provide the name of the generic function, the signature of the method (which is the type(s) used for dispatching the method), and the definition of the function.

When a generic function is called, the concrete implementation is chosen based on the type of the arguments to the function. This is what we call dynamic dispatch, and in S3 it is based on a single argument—typically the first—but in S4 we can dispatch based on more complex type information. To get behavior similar to S3 we just provide a signature that is a class name. This, then, makes S4 choose a given implementation based on the class of the first argument to the generic function.

We can implement the vector stack thus:

```
setMethod("top", signature = "VectorStack",
          definition = function(stack) stack@elements[1])
setMethod("pop", signature = "VectorStack",
          definition = function(stack) {
              VectorStack(elements = stack@elements[-1])
          })
setMethod("push", signature = "VectorStack",
          definition = function(stack, element) {
              VectorStack(elements = c(element, stack@elements))
          })
setMethod("is_empty", signature = "VectorStack",
          definition = function(stack) length(stack@elements) == 0)
```

The implementations are just variations of the functions we defined in the S3 implementation, and we can use the class just as before.

```
stack <- VectorStack()
stack <- push(stack, 1)
stack <- push(stack, 2)
stack <- push(stack, 3)
stack
## An object of class "VectorStack"
## Slot "elements":
## [1] 3 2 1
while (!is_empty(stack)) {
  stack <- pop(stack)
}
stack
## An object of class "VectorStack"
## Slot "elements":
## numeric(0)
```

Slot Prototypes

When you create an object, and you don't provide values for the slots, you will get default values, which are often empty vectors or empty lists. For example, if we create a class for representing natural numbers, we can write the class like this:

```
NaturalNumber <- setClass("NaturalNumber",
                          slots = c(
                              n = "integer"
                          ))
```

If we instantiate it without arguments the object will contain an empty vector for the natural number it is supposed to represent.

```
(n <- NaturalNumber())
## An object of class "NaturalNumber"
## Slot "n":
## integer(0)
```

If, instead, we want other default values, we can use the prototype argument to setClass. For example, we can state that the default natural number is zero.

```
NaturalNumber <- setClass("NaturalNumber",
                          slots = c(
                            n = "integer"
                          ),
                          prototype = list(
                            n = as.integer(0)
                          ))
```

Now, when we create an object, it will get the default value from the prototype.

```
(n <- NaturalNumber())
## An object of class "NaturalNumber"
## Slot "n":
## [1] 0
```

This, of course, doesn't prevent us from specifying other values as arguments when we create an object.

```
(n <- NaturalNumber(n = as.integer(1)))
## An object of class "NaturalNumber"
## Slot "n":
## [1] 1
```

Object Validity

The type we give slots when we specify them puts type constraints on objects. When we declared that the n slot in NaturalNumber should be an integer, we constrained the values we can assign to that slot. If we try to assign a numeric instead, we will get an error.

For natural numbers, we do not want negative integers to be included, but since negative integers are still integers, there is no constraint to assigning such a value.

```
n@n <- as.integer(-1)
```

We can put further constraints on objects via the `validity` argument to `setClass`. This argument should be a function that tests if an object is valid. If it is, it should return TRUE. Otherwise, it should return FALSE.

```
NaturalNumber <- setClass("NaturalNumber",
                          slots = c(
                            n = "integer"
                          ),
                          prototype = list(
                            n = as.integer(0)
                          ),
                          validity = function(object) {
                            object@n >= 0
                          })
```

With this validity checking, attempting to write the following will result in an error:

```
n <- NaturalNumber(n = as.integer(-1))
```

The validity test is only done when creating objects, though. We can modify objects and put them in an invalid state.

```
n@n <- as.integer(-1)
```

This behavior is necessary since, when modifying an object, it is likely to be in an invalid state until we are done modifying it. At any point when you are done modifying an object, though, you can call the `validObject` function to check the validity again, using the following:

```
validObject(n)
```

This would fail in this case, of course, since we just left n in an invalid state.

Generic Functions and Class Hierarchies

To see how S4 handles class hierarchies and generic functions, we return to the A, B, C example from earlier. We can construct the classes and the three objects thus:

```
A <- setClass("A", contains = "NULL")
B <- setClass("B", contains = "A")
C <- setClass("C", contains = "B")
```

```
x <- A()
y <- B()
z <- C()
```

Here I let A inherit from the pseudo-class "NULL" to make it non-abstract so that I can instantiate it even though it doesn't have any slots.

We can define a generic function f and only implement it for class A, like so:

```
setGeneric("f", def = function(x) standardGeneric("f"))
setMethod("f", signature = "A",
          definition = function(x) print("A::f"))
```

If we do, then this version will be called when we call it on all three objects.

```
f(x)
## [1] "A::f"
f(y)
## [1] "A::f"
f(z)
## [1] "A::f"
```

If we define another function, g, that we implement for both A and B, then calling it on x will call the A version. Calling it on y and z will invoke the B version since this is the most specialized form of the function for those two classes.

```
setGeneric("g", def = function(x) standardGeneric("g"))
setMethod("g", signature = "A",
          definition = function(x) print("A::g"))
setMethod("g", signature = "B",
          definition = function(x) print("B::g"))
g(x)
## [1] "A::g"
g(y)
## [1] "B::g"
g(z)
## [1] "B::g"
```

If we define a function that we implement for all three classes, then calling it on x, y, and z will invoke the most specialized version in all three cases.

```
setGeneric("h", def = function(x) standardGeneric("h"))
setMethod("h", signature = "A",
          definition = function(x) print("A::h"))
setMethod("h", signature = "B",
          definition = function(x) print("B::h"))
```

79

```
setMethod("h", signature = "C",
          definition = function(x) print("C::h"))
h(x)
## [1] "A::h"
h(y)
## [1] "B::h"
h(z)
## [1] "C::h"
```

The analogue of NextMethod in S4 is called callNextMethod and it works very similarly.

```
setMethod("h", signature = "A",
          definition = function(x) {
            print("A::h")
          })
setMethod("h", signature = "B",
          definition = function(x) {
            print("B::h")
            callNextMethod()
          })
setMethod("h", signature = "C",
          definition = function(x) {
            print("C::h")
            callNextMethod()
          })
h(x)
## [1] "A::h"
h(y)
## [1] "B::h"
## [1] "A::h"
h(z)
## [1] "C::h"
## [1] "B::h"
## [1] "A::h"
```

There is no .default version of methods as such, but we can use the setGeneric function to create one. If we define a plain old function and call setGeneric just with its name, that function will become the default function called when we do not have a more specialized version.

```
d <- function(x) print("default::d")
setGeneric("d")
d(x)
## [1] "default::d"
d(y)
## [1] "default::d"
d(z)
## [1] "default::d"
```

This, of course, also works when we specialize and use callNextMethod.

```
setMethod("d", signature = "A",
          definition = function(x) {
            print("A::d")
            callNextMethod()
          })
setMethod("d", signature = "B",
          definition = function(x) {
            print("B::d")
            callNextMethod()
          })
setMethod("d", signature = "C",
          definition = function(x) {
            print("C::d")
            callNextMethod()
          })
d(x)
## [1] "A::d"
## [1] "default::d"
d(y)
## [1] "B::d"
## [1] "A::d"
## [1] "default::d"
d(z)
## [1] "C::d"
## [1] "B::d"
## [1] "A::d"
## [1] "default::d"
```

81

Requiring Methods

The abstract class Stack we defined earlier didn't serve any purpose. It is an abstract class with no associated data or functions. All functions in the stack interface were implemented in VectorStack, and we didn't gain anything from inheriting from Stack. In S4, however, we can formalize interfaces such as stacks and ensure that implementations of an interface actually implement all the functions in the interface.

It's not much. There is very little type checking in R, and you won't get much assistance from S4 either, but there is a way of at least making the error messages more informative when you invoke a generic function that hasn't been implemented.

Let's implement a non-functioning stack. We can make this class for the list-based stack. It inherits from Stack, but it doesn't add any slots or any functionality.

```
ListStack <- setClass("ListStack", contains = "Stack")
```

Even without an implementation, we can create an object of type ListStack. The Stack class is abstract because we didn't add any slots to it, but the ListStack is not interpreted as abstract, even though it doesn't add any slots either because it contains a superclass. If we were to call pop on a ListStack, however, we would get an error, and rightly so. We would expect to get an error here, and in this case it probably isn't hard to figure out, from the error message, what is wrong. The error would tell us that R was unable to find an inherited method for the function. But if either Stack implemented a version of pop, or we had set a default function, we would instead be calling that, which would be an error but might not invoke an error message.

We can specify that all sub-classes of Stack must implement the stack interface using the function requireMethods:

```
requireMethods(functions = c("top", "pop", "push", "is_empty"),
               signature = "Stack")
```

By doing this we ensure that calling any of these methods on ListStack will give us an error message.

```
pop(stack)
## An object of class "VectorStack"
## Slot "elements":
## numeric(0)
```

It is not much of a safety check for the correct implementation of an interface, and I usually don't see much use for it, but it is there if you want it.

Constructors

You can provide values for slots when you create objects by providing them as named arguments, but you can also get more control over object initialization through the method `initialize`. This method works as a constructor, except that it doesn't create an object; it is just responsible for setting attributes to leave it in a consistent state. If you define this function, it replaces the default constructor, and you are in charge of which arguments the constructor should take, how it should set slots, and whether it should call the constructor of its superclass.

A very simple example, where we have one superclass and one subclass, is shown below:

```
A <- setClass("A", slots = list(x = "numeric", y = "numeric"))
B <- setClass("B", contains = "A", slots = list(z = "numeric"))

setMethod("initialize", signature = "A",
          definition = function(.Object, x, y) {
             print("A initialize")
             .Object@x <- x
             .Object@y <- y
             .Object
          })
## [1] "initialize"
setMethod("initialize", signature = "B",
          definition = function(.Object, z) {
             .Object <- callNextMethod(.Object, x = z, y = z)
             .Object@z <- z
             .Object
          })
## [1] "initialize"
(a <- A(x = 1:3, y = 4:6))
## [1] "A initialize"
## An object of class "A"
## Slot "x":
## [1] 1 2 3
##
## Slot "y":
## [1] 4 5 6
(b <- B(z = 6:9))
## [1] "A initialize"
## An object of class "B"
## Slot "z":
## [1] 6 7 8 9
```

```
##
## Slot "x":
## [1] 6 7 8 9
##
## Slot "y":
## [1] 6 7 8 9
```

Dispatching on Type-Signatures

The signatures we have used so far when defining specialized methods consisted of just a class name. If we use them this way, S4 methods work just as S3 generic functions, but the dispatch mechanism for S4 methods is more general than this and it is possible to dispatch based on the type of all a function's arguments.

For example, we can define a function f of two arguments and refine it in different ways based on the type of the two arguments. We could have, say, one version when the arguments are numeric and another when they are logical.

```
setGeneric("f", def = function(x, y) standardGeneric("f"))
setMethod("f", signature = c("numeric", "numeric"),
        definition = function(x, y) x + y)
setMethod("f", signature = c("logical", "logical"),
        definition = function(x, y) x & y)
```

When calling f, the appropriate function is then selected based on the type of the arguments.

```
f(2, 3)
## [1] 5
f(TRUE, FALSE)
## [1] FALSE
```

The type matching goes from most specific to most abstract, following class hierarchies for classes, and would match integer over numeric over complex for numerical values. So, if we define a version of f that matches integers for the first value, it will call that one when we give it an integer and the version defined above when we call it with numeric values.

```
setMethod("f", signature = c("integer", "complex"),
        definition = function(x, y) x - y)
f(2, 2)
## [1] 4
f(as.integer(2), 2)
## [1] 4
```

Here, the second argument would catch any complex number, but we can specialize it to match integers and numeric instead:

```
setMethod("f", signature = c("integer", "numeric"),
        definition = function(x, y) 2*x + y)
f(as.integer(2), 2)
## [1] 6
```

If we just give the signature a single string, as we did in the cases with classes earlier, it just dispatches on the type of the first argument.

```
setMethod("f", signature = "character",
        definition = function(x, y) x)
f("foo", "bar")
## [1] "foo"
```

In general, the signature list just matches types for a prefix of parameters if you do not provide types for all of them.

```
setGeneric("g", def = function(x, y, z) standardGeneric("g"))
setMethod("g", signature = "character",
        definition = function(x, y, z) "g(character)")
setMethod("g", signature = c("numeric", "character"),
        definition = function(x, y, z) "g(numeric, character)")
g("foo", NA, NA)
## [1] "g(character)"
g(12, "bar", NA)
## [1] "g(numeric, character)"
```

If you want to match any type whatsoever, you can use the type "ANY".

```
setMethod("f", signature = "ANY",
        definition = function(x, y) "any")
f(list(), NULL)
## [1] "any"
```

This can be used to define a catch-all default implementation for when no more specific implementation matches the arguments.

You can even match for cases when some arguments are not provided, using the type "missing".

```
setMethod("f", signature = c("ANY", "missing"),
          definition = function(x, y) "missing")
f(list(), NULL)
## [1] "any"
f(list())
## [1] "missing"
```

Here, the first call matches the version that takes any arguments because the second argument is not missing, it is just NULL, while the other matches the more specific signature where the second argument is missing.

Operator Overloading

S4 also supports operator overloading and in much the same way as S3 does, just using the method mechanism for generic methods. We can try implementing the modulus class as an S4 class like this:

```
modulus <- setClass("modulus",
                    slots = c(
                      value = "numeric",
                      n = "numeric"
                    ))

setMethod("show", signature = "modulus",
          definition = function(object) {
            cat("Modulus", object@n, "values:\n")
            print(object@value)
          })
(x <- modulus(value = 1:6, n = 3))
## Modulus 3 values:
## [1] 1 2 3 4 5 6
```

The show method we implemented here is the S4 equivalent of print and we use it to pretty-print modulus objects.

If we then want to implement a single operator, we can specialize the method for it, just as we did with generic functions for S3. But we can use the signature type matching to capture different combinations of arguments, instead of writing type-checking code at the beginning of the generic function as we had to for S3.

Below we handle the three cases we want for modulus arithmetic: the case where both operands are modulus objects and the two cases where one of them is a modulus object and the other is numeric.

```
setMethod("+", signature = c("modulus", "modulus"),
          definition = function(e1, e2) {
            if (e1@n != e2@n) stop("Incompatible modulus")
            modulus(value = e1@value + e2@value,
                    n = e1@n)
          })
setMethod("+", signature = c("modulus", "numeric"),
          definition = function(e1, e2) {
            modulus(value = e1@value + e2,
                    n = e1@n)
          })
setMethod("+", signature = c("numeric", "modulus"),
          definition = function(e1, e2) {
            modulus(value = e1 + e2@value,
                    n = e2@n)
          })
```

Now we can combine numeric and modulus in addition, and we didn't have to explicitly check the type in the functions since the type dispatch handled that for us.

```
x + 1:6
## Modulus 3 values:
## [1]  2  4  6  8 10 12
1:6 + x
## Modulus 3 values:
## [1]  2  4  6  8 10 12
```

We also handle the case with two modulus objects and check that their n slots are equal.

```
y <- modulus(value = 1:6, n = 2)
x + y
## Error in x + y: Incompatible modulus
y <- modulus(value = 1:6, n = 3)
x + y
## Modulus 3 values:
## [1]  2  4  6  8 10 12
```

We also have function group solutions in S4, and for defining arithmetic operations, we need the group Arith. This works much as the Ops generic function in S3 except that we, again, can use type matching instead of explicitly checking the type of the arguments inside the function(s).

```
setMethod("Arith",
          signature = c("modulus", "modulus"),
          definition = function(e1, e2) {
            if (e1@n != e2@n) stop("Incompatible modulus")
            modulus(value = callGeneric(e1@value, e2@value),
                    n = e1@n)
          })
setMethod("Arith",
          signature = c("modulus", "numeric"),
          definition = function(e1, e2) {
            modulus(value = callGeneric(e1@value, e2),
                    n = e1@n)
          })
setMethod("Arith",
          signature = c("numeric", "modulus"),
          definition = function(e1, e2) {
            modulus(value = callGeneric(e1, e2@value),
                    n = e2@n)
          })
x * y
## Modulus 3 values:
## [1]  1  4  9 16 25 36
2 * x
## Modulus 3 values:
## [1]  2  4  6  8 10 12
```

Combining S3 and S4 Classes

You can, to a limited degree, combine S3 and S4. The two systems are different and trying to write software that connects S3 and S4 class-hierarchies intimately is not something I will recommend. It only leads to weeping and gnashing of teeth. But if you have existing S3 code and you want to write an extension in S4 you can do this.

Let's say we have an S3 class, X, with generic functions foo and bar.

```
X <- function(x) {
  structure(list(x = x), class = "X")
}

foo <- function(x) UseMethod("foo")
bar <- function(x) UseMethod("bar")
foo.X <- function(x) "foo"
bar.X <- function(x) x$x
```

```
x <- X(5)
foo(x)
## [1] "foo"
bar(x)
## [1] 5
```

If we want to write a subclass of X, let's call it Y, and we want to write Y in S4, we cannot use X in the contains option to setClass. Well, you can, but if you try to instantiate objects of the class you will get an error. The S4 system doesn't know about any class called X so we first have to make it aware of it. We can do that using the function setOldClass.

```
setOldClass("X")
```

This does two things: it lets S4 know about the class so we can inherit from it, and it makes any generic function defined for the class into functions we can specialize with setMethod. So after calling setOldClass we can make a sub-class of A as an S4 class.

```
Y <- setClass("Y", contains = "X")
```

Calling foo or bar on an object of type Y will also work with the dynamic dispatch system and will invoke foo.Y since there is no better matching foo.Y.

```
y <- Y()
foo(y)
## [1] "foo"
```

If we want to, we could make a specialized version of foo for Y objects by implementing foo.Y.

```
foo.Y <- function(x) "Y::foo"
foo(y)
## [1] "Y::foo"
```

Of course, this would be the S3 of refining generic functions, and since we are working with an S4 class now it is better to use setMethod.

Similar to foo, calling bar invokes bar.X. In this case it results in an error because, even though S4 knows about the X class, it doesn't know about the constructor function or the representation of X objects it creates.

```
bar(y)
## Error in x$x: $ operator not defined for this S4 class
```

There is no formal definition of constructors or object representations in S3, only informal coding conventions, so no way for S4 to know about what the bar function expects to be able to get out of an X object. There is a limit to how well we can integrate S3 and S4 automatically, and some coding is needed to get the functionality of the S3 version to also match the S4 class.

```
Y <- setClass("Y", contains = "X", slots = c(x = "ANY"))
setMethod("bar", signature = "Y",
          definition = function(x) x@x)
y <- Y(x = 13)
bar(y)
## [1] 13
```

I don't recommend mixing S3 and S4. If you have code written using the S3 system you are probably better off sticking with S3 rather than trying to combine the two systems, but if you are writing code using S4 and need to include a little functionality from S3 classes, this is the way to do it.

## CHAPTER 7

R6 Classes

The last object system we will look at is the R6 system. This system is very unlike S3 and S4, and unusual for R in general, since it introduces mutable data. Data is usually immutable in R, and anything resembling object modification is really implemented by copying data and constructing new objects. The only things you can modify in R are environments; you cannot modify data. The R6 system, however, uses environments to give us objects we can modify. It's unusual in R, but the semantics are then similar to how object-orientation is implemented in most other languages, where methods modify objects rather than create new objects. If you come to R from a different programming language, the R6 system might be more familiar to you, but since it introduces side effects to R, it might surprise you if you are more familiar with R. If you have two references to an R6 object and modify it through one variable, it will also change the state of the object the other variable is referring to. This does not happen for the other two class systems.

The R6 system is a better implementation of this semantics than the built-in reference class (RC) system, also known as R5. R5 is the natural name for the next object system implemented in R, after the object systems S3 and S4, which are originally from the S language. Because R6 has the same semantics as R5 and is considered a better implementation of it, I will not cover R5 further.

Defining Classes

Classes are defined using the R6Class function from the package R6. Similar to setClass from S4, we need to give the class a name, and we have a number of optional arguments for defining how objects of the class should look and behave. Unlike S4, we need to capture the result from the call to R6Class in order to create objects of the class. In S4 this is a convention, but we can create objects as long as we know the name of a class. In R6, the name is mainly used to set the class attribute so the objects can interact with S3 polymorphism. It is the return value of R6Class we use to create objects.

As an example, once again we implement a stack. This time I will not bother with an abstract super class but just implement a VectorStack directly. One implementation can look like this:

```
library(R6)

VectorStack <- R6Class("VectorStack",
                    private = list(elements = NULL),
                    public = list(
                      top = function() {
                        private$elements[1]
                      },
                      pop = function() {
                        private$elements <-
                          private$elements[-1]
                        invisible(self)
                      },
                      push = function(e) {
                        private$elements <-
                          c(e, private$elements)
                        invisible(self)
                      },
                      is_empty = function() {
                        length(private$elements) == 0
                      }
                    ))
```

Here, besides the class name, we use two arguments to R6Class: private and public. These are used to define attributes of the class, either values stored in objects or methods we can call on the objects. For both, the arguments are lists. The names used for the list elements become the names of the attributes and the values—naturally, the attribute values.

The difference between the two arguments is that attributes in the public list can be accessed anywhere while attributes in the private list can only be accessed in methods you define for the class. In methods, you can access elements in the public list using the variable self, and you can access attributes in the private list using the variable private. In this VectorStack implementation, we have made the vector used for storing the stack private, and we have implemented the stack interface as public methods. Inside the methods we access the elements as private$elements, and in push and pop we return the object itself using the variable self. We return this object wrapped in invisible, so it isn't automatically printed when we call these methods, but we didn't have to. We didn't have to return an object at all for these methods, but doing so allows us to chain together method calls, as we will see below, so it is good practice.

Notice that the self and private objects are not arguments to the methods. They just exist in the namespace of the functions as part of the magic R6 uses to implement its mutable object semantics.

The VectorStack object we create this way is not a constructor function itself, as it was for S4. It is a so-called object generator. To create an object we use the attribute new of this generator thus:

```
(stack <- VectorStack$new())
## <VectorStack>
##   Public:
##     clone: function (deep = FALSE)
##     is_empty: function ()
##     pop: function ()
##     push: function (e)
##     top: function ()
##   Private:
##     elements: NULL
```

Printing this object is rather verbose, but if we define the method print for the class we can modify how it is displayed.

```
VectorStack <- R6Class("VectorStack",
                       private = list(elements = NULL),
                       public = list(
                         # ... rest of the methods
                         print = function() {
                           cat("Stack elements:\n")
                           print(private$elements)
                         }
                       ))
```

This doesn't modify the existing object.

```
stack
## <VectorStack>
##   Public:
##     clone: function (deep = FALSE)
##     is_empty: function ()
##     pop: function ()
##     push: function (e)
##     top: function ()
##   Private:
##     elements: NULL
```

93

It has an effect if we create a new one, however.

```
(stack <- VectorStack$new())
## Stack elements:
## NULL
```

We can access the (public) attributes of the stack object, and call methods if the attributes are functions, using $ indexing:

```
stack$push(1)$push(2)$push(3)
stack
## Stack elements:
## [1] 3 2 1
while (!stack$is_empty()) stack$pop()
stack
## Stack elements:
## numeric(0)
```

The chained call to push here is possible because the push method returns the object itself. Unlike the previous implementations where push returns a new object, for the R6 object, the existing object is modified. We do not need to assign the result of the three push calls back to stack, and we do not need to assign the calls to pop back to stack either. Returning the object itself in these functions allows us to chain method calls, but that is all this does. The R6 object is not immutable.

Object Initialization

If we want to set attributes of objects when constructing them, we need to do a little more work than in S4. We cannot just use named arguments in the constructor; this call would give us an error:

```
stack <- VectorStack$new(elements = 1:4)
```

To be able to initialize objects this way, we need to explicitly write a function for it. This function must be a public method called initialize. If you want the constructor to take arguments, you must specify the arguments in this function. You cannot make this function private; it is an error to put a function named initialize in the private list.

To initialize VectorStack objects with a sequence of elements we can implement its initialize function like this:

```
VectorStack <- R6Class("VectorStack",
                    private = list(elements = NULL),
                    public = list(
                      initialize = function(elements = NULL) {
                        private$elements <- elements
                      },
                      # ... rest of the methods
                    ))
```

With this initialization function we can now construct objects with elements initialized.

```
(stack <- VectorStack$new(elements = 1:4))
## Stack elements:
## [1] 1 2 3 4
```

Private and Public Attributes

The elements in the stack are private, so we cannot access them the same way we can the public methods. You might hope that stack$elements would then give you an error, but unfortunately not.

```
stack$elements
## NULL
```

This is because accessing a list with a key it doesn't contain gives you NULL and it is this behavior you are getting.

```
list()$elements
## NULL
```

We can see the difference between private and public attributes with this little example:

```
A <- R6Class("A", public = list(x = 5), private = list(y = 13))
```

With this definition of class A, we should be able to access object's x attributes, and we can.

```
a <- A$new()
a$x
## [1] 5
a$x <- 7
a$x
## [1] 7
```

We can also get y, but it has the behavior we saw above for stacks, and we are not allowed to modify it since it isn't really an attribute of the object.

```
a$y
## NULL
a$y <- 12
## Error in a$y <- 12: cannot add bindings to a locked environment
a$y
## NULL
```

In general, you cannot create new attributes to R6 objects just by assigning to $ indexed values, as you can in S3. Attributes must be defined in the class definition. If you tried something like this, you would get an error:

```
a$z <- "foo"
```

You can modify public data attributes, as we saw above for x, but don't try to be clever and modify methods. It is really bad practice to change methods for a single object to begin with, but luckily it is also "verboten" in R6, and you would get you an error if you tried.

In general, it is considered good practice to keep data private and methods that are part of a class interface public. There are several reasons for this: if data is only modified through a class's methods then you have more control over the state of objects and can ensure that an object is always in a valid state before and after all method calls, but (perhaps more importantly) keeping the representation of objects hidden away limits the dependency between a class and code that uses the class. If any code can access the inner workings of objects, there is a good chance that eventually a lot of code will. This means that you will have to modify all the uses of a class if you change how objects of the class are represented. If on the other hand, the code only accesses objects through a public interface, then you can modify all the private attributes as much as you want as long as you keep the public interface unchanged. You will, of course, have to modify some of the class's methods, but changes will be limited to that.

In the R6 system, private attributes can be accessed only by methods you define for the class or in methods defined in sub-classes. If you are used to languages such as C++ or Java, this might surprise you, but the private attributes in R6 are similar to the protected attributes in those languages and not the private attributes.

Active Bindings

There is a way of getting the syntax of accessing data attributes without actually doing so. If you have code that already uses a public attribute and you want to change that into a function to hide or modify implementation details, you can use this. You can also use it if you just like the syntax for data better than method calls.

This is achieved through the `active` argument to R6Class. Here you can provide a list of attributes, as for `private` and `public`, but these attributes should be functions, and they will define a value-like syntax for calling the functions.

As an example, we can take the elements in the vector stack. We want to be able to write `stack$elements`, but we do not want to make the elements public. So we write a function for `elements` and add it to `active`. We cannot have the same name used both in `private` and `active` (or in `public` for that matter), so we have to change the name for the private data attribute first, and of course update all the existing methods. After doing that, we can add the `active` function like this:

```
VectorStack <- R6Class("VectorStack",
                    private = list(elements_ = NULL),
                    public = list(
                      # ... methods
                    ),
                    active = list(
                      elements = function(value) {
                        if (!missing(value))
                          stop("elements are read-only")
                        private$elements_
                      }
                    ))
```

Functions in the `active` list should take one argument, `value`. This value will be missing when we read the attribute, and it will contain data when we assign to the attribute. In this implementation, we consider assigning to the elements an error and we return the private `elements_` when we read the attribute.

This will give us the elements:

```
stack <- VectorStack$new(elements = 1:4)
stack$elements
## [1] 1 2 3 4
```

Meanwhile this will raise an error:

```
stack$elements <- rev(1:3)
## Error in (function (value) : elements are read-only
```

You can use these active functions to modify values you assign, whether to ensure object consistency, or to fake an attribute that isn't directly stored but exists implicitly by being computable from other data. It all depends on how you choose to use them.

Inheritance

The way we specify class hierarchies, and the way method calls are dispatched to the most specialized implementation of a method, is relatively straightforward. We can take the example of three classes we have seen two times earlier and implement it in R6. To specify that one class inherits from another, we use the inherit argument to R6Class, and to write a more specialized version of a method, we simply add the method to the public or private lists. Overall, writing methods and class hierarchies is done with much simpler code in R6 than in both S3 and S4.

```
A <- R6Class("A",
            public = list(
                f = function() print("A::f"),
                g = function() print("A::g"),
                h = function() print("A::h")
            ))
B <- R6Class("B", inherit = A,
            public = list(
                g = function() print("B::g"),
                h = function() print("B::h")
            ))
C <- R6Class("C", inherit = B,
            public = list(
                h = function() print("C::h")
            ))
```

There are no surprises in how we instantiate objects of the classes; we have to use the new method in the object generators.

```
x <- A$new()
y <- B$new()
z <- C$new()
```

For method f we only have an implementation for class A, so calling f on all three objects will call that version. Except that the method call has a different syntax from the implementations for S3 and S4, there are no surprises here.

```
x$f()
## [1] "A::f"
y$f()
## [1] "A::f"
z$f()
## [1] "A::f"
```

For g, we have implementations in both A and B, and the C object will call the B implementation since this is the most specialized for that class.

```
x$g()
## [1] "A::g"
y$g()
## [1] "B::g"
z$g()
## [1] "B::g"
```

Finally, for h we have implementations in all three classes, so we call different methods for the three objects.

```
x$h()
## [1] "A::h"
y$h()
## [1] "B::h"
z$h()
## [1] "C::h"
```

There should not be any surprises in how inheritance and method dispatching works in R6.

References to Objects and Object Sharing

One important thing is different from R6 objects and all other R objects: the R6 objects have a state that can be modified. If you are used to other object-oriented programming languages, this might not sound like much of a deal, but in general, we can assume that calling functions do not have side effects in R, except for changing values that variables point to. When objects can suddenly change state, we need to worry about when two references are to the same object or merely references to two objects that represent the same values.

The first thing you need to know is that values set in the definition of private and public lists are shared between objects of a class. To see this in action we can define these two classes:

```
A <- R6Class("A", public = list(x = 1:5))
B <- R6Class("B",
                public = list(
                x = 1:5,
                a = A$new()
                ))
```

Here, I am breaking the rule about not having public data to simplify the example. In any case, what we have is one class, A, that contains a vector, and another, B, that contains another vector and a reference to an A object. Let's create two objects of class B.

```
x <- B$new()
y <- B$new()
```

We can first check the behavior of the vector in the objects. It is initialed to the first five natural numbers, so that is what both objects contain initially.

```
x$x
## [1] 1 2 3 4 5
y$x
## [1] 1 2 3 4 5
```

If we then modify the vector in x we see that this vector changes but the vector in y does not. This is how vectors behave in R and generally what we would expect.

```
x$x <- 1:3
x$x
## [1] 1 2 3
y$x
## [1] 1 2 3 4 5
```

If we modify the vector in the nested A object, however, we get a different behavior. Here, changing the value through x *also* changes the value in y.

```
x$a$x
## [1] 1 2 3 4 5
y$a$x
## [1] 1 2 3 4 5
```

```
x$a$x <- 1:3
x$a$x
## [1] 1 2 3
y$a$x
## [1] 1 2 3
```

Even creating a new object from the class will give us an object that contains the modified value.

```
z <- B$new()
z$a$x
## [1] 1 2 3
```

All three objects are referring to the same A object and modifications to this object are reflected in all of them. This is generally how R6 classes behave. The copy-on-modification semantics of other R objects are not how R6 objects behave. When you have two references to the same object then modifying one of them will also modify the other.

Modifying x$x didn't change y$x because x and y are different objects, but if we make another reference to the object pointed to by x then changes to x will be reflected in the other.

```
w <- x
w$x
## [1] 1 2 3
x$x <- 1:5
w$x
## [1] 1 2 3 4 5
```

If you want each object of a class to contain distinct objects of an R6 class, then you can create the objects in the `initialize` function instead of in the `public` or `private` lists. This function is called whenever you create a new object and objects that are created in the initialization function will be distinct and thus not shared.

We can modify B like this:

```
B <- R6Class("B",
            public = list(
              x = 1:5,
              a = NULL,
              initialize = function() {
                self$a <- A$new()
              }))
```

We need to re-create x and y to have them refer to this new class:

```
x <- B$new()
y <- B$new()
```

Now we can modify one without modifying the other.

```
x$a$x
## [1] 1 2 3 4 5
x$a$x <- 1:3
x$a$x
## [1] 1 2 3
y$a$x
## [1] 1 2 3 4 5
```

Since assigning from one variable to another just create another reference to the same object, we need another way of creating an effective copy. This is done with the clone method that all R6 objects automatically implement.

If we clone object x we get a new copy of the object, which contains the same state as x does at the time of cloning, but which can be modified without changing x.

```
z <- x$clone()
z$x
## [1] 1 2 3 4 5
z$x <- 1:2
x$x
## [1] 1 2 3 4 5
```

The default cloning is shallow, however. It makes a copy of the object, but if the object contains a reference to an R6 class, then the clone will contain a reference to the same object. If we modify the a attribute of z, we will also modify the a attribute of x.

```
x$a$x
## [1] 1 2 3
z$a$x <- 1:5
x$a$x
## [1] 1 2 3 4 5
```

If we call clone with the option deep = TRUE we will instead get a deep copy; here we get a transitive closure of cloned references, so here we can modify the a attribute safe in the knowledge that they are distinct between an object and its clone.

```
y <- x$clone(deep = TRUE)

x$a$x
## [1] 1 2 3 4 5
y$a$x <- NULL
x$a$x
## [1] 1 2 3 4 5
```

Interaction with S3 and Operator Overloading

We don't have a mechanism for defining new operators for R6 objects, but we
can use the S3 system for this. Objects create from R6 object generators are
assigned a class attribute, a list of the name we give the class when creating the
generator and "R6", so we can define generic function specializations for them.

We can implement the modulus class in R6 like this:

```
modulus <- R6Class("modulus",
                private = list(
                  value_ = c(),
                  n_ = c()
                ),
                public = list(
                  initialize = function(value, n) {
                    private$value_ <- value
                    private$n_ <- n
                  },
                  print = function() {
                    cat("Modulus", private$n_, "values:\n")
                    print(private$value_)
                  }
                ),
                active = list(
                  value = function(value) {
                    if (missing(value)) private$value_
                    else private$value_ <- value %% private$n_
                  },
                  n = function(value) {
                    if (!missing(value)) stop("Cannot change n")
                    private$n_
                  }
                ))
```

```
(x <- modulus$new(value = 1:6, n = 3))
## Modulus 3 values:
## [1] 1 2 3 4 5 6
```

There are a few things going on in this class definition. We define the attributes for holding the data in the `private` list, define an initialization function and a print function, and then we define two `action` attributes for accessing the data. We allow users of the class to modify `values` but not `n` (for no good reason other than it gives us an example of two different behaviors), and for the `values` attribute we make sure that we modify the data before we store it in the private `values_` variable.

The `class` attribute of objects of this class contains this:

```
class(x)
## [1] "modulus" "R6"
```

This means that we can define arithmetic operations on the class using the S3 system like this:

```
Ops.modulus <- function(e1, e2) {
  nx <- ny <- NULL
  if (inherits(e1, "modulus")) nx <- e1$n
  if (inherits(e2, "modulus")) ny <- e2$n
  if (!is.null(nx) && !is.null(ny) && nx != ny)
    stop("Incompatible types")
  n <- ifelse(!is.null(nx), nx, ny)

  v1 <- e1
  v2 <- e2
  if (inherits(e1, "modulus")) v1 <- e1$value
  if (inherits(e2, "modulus")) v2 <- e2$value

  e1 <- v1 ; e2 <- v2
  result <- NextMethod() %% n
  modulus$new(result, n)
}
```

The implementation is slightly different from the S3 version, because the data in the objects are represented differently, but the general control flow is the same, and with this definition we have modulus arithmetic.

```
x + 1:6
## Modulus 3 values:
## [1] 2 1 0 2 1 0
```

```
1:6 + x
## Modulus 3 values:
## [1] 2 1 0 2 1 0
2 * x
## Modulus 3 values:
## [1] 2 1 0 2 1 0
```

If you make subclasses of an R6 class like modulus, you will get a class attribute that also reflects this, so the S3 dispatch mechanism will also work for sub-classes in the R6 system.

```
modulus2 <- R6Class("modulus2", inherit = modulus)
y <- modulus2$new(value = 1:2, n = 3)
class(y)
## [1] "modulus2" "modulus"  "R6"
x + y
## Modulus 3 values:
## [1] 2 1 1 0 0 2
```

That being said, don't go crazy with combining R6 and S3 either; it will only confuse the maintainers of your code (which are likely to include yourself sometime in the future).

CHAPTER 8

Conclusions

This concludes this book on object-oriented programming in R. You now know the three different systems for object-oriented programming in R and how to use them to define class hierarchies and polymorphic functions.

Object-oriented programming in R, at least for the S3 and S4 system, differs from most other object-oriented programming languages. Most languages consider objects mutable, and most object-oriented software designs involve wiring up objects with references to each other such that their behavior depends on the changing states of other objects. The R6 system is closer to this type of language design. Still, the S3 and S4 systems combine two powerful programming language paradigms: functional programming and object-oriented programming. The combination of dynamic function dispatch based on the argument types and high-level functional programming lets you construct flexible and extensible software.

It can be confusing with three very different systems for object-oriented programming in the same language, and I would recommend that you stick to one for any single project. Knowing all three, however, and knowing the pros and cons of using them lets you pick the right tool for any particular job. The S3 system is the simplest of the three and useful for getting a small model up and running in short time. The more formal classes of S4 makes it easier to structure more complex frameworks, and the reference semantics of R6 make it simpler to implement traditional mutable data structures than you can otherwise easily do in R.

Getting familiar with these systems, of course, requires practice and you will not be an expert object-oriented programming just from reading this book. You know enough now, though, to get started.

© Thomas Mailund 2017

T. Mailund, *Advanced Object-Oriented Programming in R*, DOI 10.1007/978-1-4842-2919-4_8

Index

A

Algorithms, implementations, 18–19

B

Bayesian linear regression, 43–47

C, D, E

Classes, 3–5
Class hierarchies
    abstract and concrete classes, 23–25
    bibliography objects, 31–33
    generic functions, 30
    GraphicalObject, 25–29
    implementation, 22
    interface, 21–22
    polymorphism, 22
    prune function, 30, 31
Composite design pattern, 34
Constructors, 83–84

F

Fitted model objects, 52–54

G, H

Generic functions
    stack, 1
    UseMethod, 2
GraphicalObject, 25–29

I, J, K, L

Interfaces, 9–12

M

Model matrices, 47–51

N

NextMethod, 39–42

O

Operator overloading, 103–105
    arithmetic modulus, 62
    group operators, 64–66
    S4 class, 86–88
    single operators, 63–64
    units, 67–71

P, Q

Polymorphic functions, 1
Polymorphism, 5–9, 22
    algorithmic
        programming, 13
    sorting lists, 14–18
    use case, 12–13
Predicting response
        variables, 54–60
Prune function, 30, 31

© Thomas Mailund 2017

T. Mailund, *Advanced Object-Oriented Programming in R*, DOI 10.1007/978-1-4842-2919-4

■ R

R6 system
    active bindings, 97
    inheritance, 98–99
    object initialization, 94–95
    objects and object
        sharing, 99–102
    private and public attributes,
        95–96
    S3 and operator overloading,
        103–105
    VectorStack, 92–94
Reuse code
    class hierarchies, 36–38
    constructors, 38–39
    generic functions, 35–36
    NextMethod, 39–42

■ S

S3 class, 103–105
    defining, 73–74
    generic functions, 75–76

hierarchies and generic functions
    callNextMethod, 80–81
    defining function, 79
    objects, 78
    requireMethods, 82
    stack, 82
object validity, 77–78
and s4, 88–90
slot prototypes, 76–77
Statistical models
    Bayesian linear regression, 43–47
    coefficients and predict, 43
    fitted model objects, 52–54
    model matrices, 47–51
    predicting response variables, 54–60

■ T

Type-signatures, 84–86

■ U, V, W, X, Y, Z

Units, 67–71
UseMethod function, 19–20

Get the eBook for only $5!

Why limit yourself?

With most of our titles available in both PDF and ePUB format, you can access your content wherever and however you wish—on your PC, phone, tablet, or reader.

Since you've purchased this print book, we are happy to offer you the eBook for just $5.

To learn more, go to http://www.apress.com/companion or contact support@apress.com.

Apress®

Printed in the United States
By Bookmasters